Methods in Molecular Biology

For further volumes:
http://www.springer.com/series/7651

For over 35 years, biological scientists have come to rely on the research protocols and methodologies in the critically acclaimed *Methods in Molecular Biology* series. The series was the first to introduce the step-by-step protocols approach that has become the standard in all biomedical protocol publishing. Each protocol is provided in readily-reproducible step-by-step fashion, opening with an introductory overview, a list of the materials and reagents needed to complete the experiment, and followed by a detailed procedure that is supported with a helpful notes section offering tips and tricks of the trade as well as troubleshooting advice. These hallmark features were introduced by series editor Dr. John Walker and constitute the key ingredient in each and every volume of the *Methods in Molecular Biology* series. Tested and trusted, comprehensive and reliable, all protocols from the series are indexed in PubMed.

Plant Immunity

Methods and Protocols

Edited by

Tsutomu Kawasaki

Graduate School of Agriculture, Kindai University, Nara, Japan

Hirofumi Yoshioka

Graduate School of Bioagricultural Sciences, Nagoya University, Nagoya, Japan

Humana Press

Editors
Tsutomu Kawasaki
Graduate School of Agriculture
Kindai University
Nara, Japan

Hirofumi Yoshioka
Graduate School of Bioagricultural Sciences
Nagoya University
Nagoya, Japan

ISSN 1064-3745 ISSN 1940-6029 (electronic)
ISBN 978-1-0716-5137-7 ISBN 978-1-0716-5138-4 (eBook)
https://doi.org/10.1007/978-1-0716-5138-4

This Humana imprint is published by the registered company Springer Science+Business Media, LLC, part of Springer Nature.
The registered company address is: 1 New York Plaza, New York, NY 10004, U.S.A.

Preface

A variety of plant immune responses are induced through the recognition of pathogen-derived molecules by immune receptors. Plasma membrane-localized receptors are classified into two types: receptor-like kinases (RLKs) and receptor-like proteins (RLPs), which recognize microbe-, pathogen-, or danger-associated molecular patterns (MAMPs/PAMPs/DAMPs). These receptors are collectively termed pattern-recognition receptors (PRRs). Although plants possess a large number of RLKs and RLPs, only a limited number of PRR–ligand pairs have been identified. Recent progress indicates that PRRs interact with multiple positive and negative immune regulators, both in the absence and presence of MAMPs/PAMPs/DAMPs. Perception of these signals by PRRs triggers Ca^{2+} influx into the cytoplasm, the production of reactive oxygen species (ROS), and MAPK activation, leading to the induction of diverse immune responses collectively termed pattern-triggered immunity (PTI). This volume provides step-by-step protocols for recently established genetic and biochemical techniques to identify PRRs and their associated downstream immune regulators. Furthermore, it covers PRR–ligand interaction-based immune responses, such as cellular Ca^{2+} dynamics, spatiotemporal imaging of CDPK activity, and ROS production by NADPH oxidase.

Successful pathogens deliver various effectors into plant cells to overcome PTI or to modify nutrient conditions for pathogen proliferation. As a second layer of immune response, intracellular nucleotide-binding leucine-rich repeat receptors (NLRs) recognize these effectors and trigger robust immune responses, often accompanied by localized cell death, termed effector-triggered immunity (ETI). Bioinformatic studies have revealed that each plant genome encodes many NLRs. Some NLRs form higher-order complexes, known as resistosomes, which are predicted to function as Ca^{2+}-permeable channels. This book provides detailed methods for pan-NLRome analysis, identification of NLR complex components, resistosome assays, and cell death evaluation.

Pathogen infection and immune responses also involve dynamic transcriptional and metabolic reprogramming in both plants and pathogens. To address the spatiotemporal changes of these processes, this volume includes protocols for co-transcriptome analysis of plant–pathogen interactions, single-nucleus multi-omics, metabolite imaging, and bioluminescence-based quantification of pathogen growth. It also describes genome editing tools that contribute to the development of disease-resistant plants.

Finally, we are deeply grateful to all the authors who contributed chapters to *Plant Immunity: Methods and Protocols*. They provided detailed protocols and valuable trouble-shooting advice. We also sincerely acknowledge Professor John M. Walker for inviting us to edit this volume and for his unfailing support. We hope this book will accelerate your ongoing research and inspire further advances in the field of plant immunity.

Nara, Japan — *Tsutomu Kawasaki*
Nagoya, Japan — *Hirofumi Yoshioka*

Contents

Contributors

AKIRA ABE • *Iwate Biotechnology Research Center, Kitakami, Japan*

HIROAKI ADACHI • *Laboratory of Crop Evolution, Graduate School of Agriculture, Kyoto University, Muko, Kyoto, Japan; Laboratory of Plant Immunity, Faculty of Advanced Life Science, Hokkaido University, Sapporo, Hokkaido, Japan*

SHIGEYUKI BETSUYAKU • *Faculty of Agriculture, Ryukoku University, 1-5 Yokotani, Seta Oe-cho, Otsu, Shiga, Japan*

SAVITHRAMMA P. DINESH-KUMAR • *Department of Plant Biology and The Genome Center, College of Biological Sciences, University of California, Davis, Davis, CA, USA*

BERNADETTE EICHSTÄDT • *Department for Biochemistry of Plant Interactions, Leibniz Institute of Plant Biochemistry, Halle (Saale), Germany*

YUGA FUJINAWA • *Laboratory of Plant Pathology, Graduate School of Agriculture, Kyoto University, Kyoto, Japan*

TADASHI FUJIWARA • *Laboratory of Crop Evolution, Graduate School of Agriculture, Kyoto University, Muko, Kyoto, Japan; Laboratory of Plant Immunity, Faculty of Advanced Life Science, Hokkaido University, Sapporo, Hokkaido, Japan*

ROBIN HON MING GOH • *Department of Cell & Systems Biology, University of Toronto, Toronto, ON, Canada*

PING HE • *Department of Molecular, Cellular, and Developmental Biology, University of Michigan, Ann Arbor, MI, USA*

YUTA HINO • *Laboratory of Plant Immunology, Graduate School of Bioagricultural Sciences, Nagoya University, Chikusa, Nagoya, Aichi, Japan*

KODAI HONDA • *Laboratory of Crop Evolution, Graduate School of Agriculture, Kyoto University, Muko, Kyoto, Japan*

ZHIYUAN JI • *State Key Laboratory of Crop Gene Resources and Breeding/National Key Facility for Crop Gene Resources and Genetic Improvement, Institute of Crop Sciences, Chinese Academy of Agricultural Sciences, Beijing, China*

JONATHAN D. G. JONES • *The Sainsbury Laboratory, University of East Anglia, Norwich Research Park, Norwich, UK*

YASUHIRO KADOTA • *RIKEN Center for Sustainable Resource Science (CSRS), RIKEN-TRIP, Yokohama, Kanagawa, Japan*

HIROAKI KATO • *Laboratory of Crop Evolution, Graduate School of Agriculture, Kyoto University, Kyoto, Japan; JST-PRESTO, Kawaguchi, Japan*

TSUTOMU KAWASAKI • *Graduate School of Agriculture, Kindai University, Nara, Japan*

SUNG-IL KIM • *Department of Molecular, Cellular, and Developmental Biology, University of Michigan, Ann Arbor, MI, USA*

SARAH LEDERER • *Department for Biochemistry of Plant Interactions, Leibniz Institute of Plant Biochemistry, Halle (Saale), Germany*

HYUNSUH LEE • *Department of Cell & Systems Biology, University of Toronto, Toronto, ON, Canada*

YUEXIAO LI • *State Key Laboratory of Plant Environmental Resilience, College of Biological Sciences, China Agricultural University, Beijing, China*

ANJA LIESE • *Department for Biochemistry of Plant Interactions, Leibniz Institute of Plant Biochemistry, Halle (Saale), Germany*

XIAO LIN • *Department of Agri-microbiomics and Biotechnology, State Key Laboratory of Microbial Diversity and Innovative Utilization, Institute of Microbiology, Chinese Academy of Sciences, Beijing, China; University of the Chinese Academy of Sciences, Beijing, China*

YAPEI LIU • *State Key Laboratory of Crop Gene Resources and Breeding/National Key Facility for Crop Gene Resources and Genetic Improvement, Institute of Crop Sciences, Chinese Academy of Agricultural Sciences, Beijing, China*

AKIRA MINE • *Laboratory of Plant Pathology, Graduate School of Agriculture, Kyoto University, Kyoto, Japan*

WOLFGANG MOEDER • *Department of Cell & Systems Biology, University of Toronto, Toronto, ON, Canada*

HIROFUMI NAKAGAMI • *Basic Immune System of Plants, Max Planck Institute for Plant Breeding Research, Cologne, Germany*

KAORU NAKAGAWA • *Shimadzu Corporation, Nishinokyo Kuwabara-cho, Nakagyo-ku, Kyoto, Japan*

TATSUYA NOBORI • *The Sainsbury Laboratory, University of East Anglia, Norwich, UK*

KOJI OKUDA • *Shimadzu Corporation, Nishinokyo Kuwabara-cho, Nakagyo-ku, Kyoto, Japan*

STEFANIE RANF • *Department of Biology, University of Fribourg, Fribourg, Switzerland*

TINA ROMEIS • *Department for Biochemistry of Plant Interactions, Leibniz Institute of Plant Biochemistry, Halle (Saale), Germany; Biochemistry of Plant Interaction, Martin-Luther-University Halle-Wittenberg, Halle (Saale), Germany*

MUHAMMAD SOHAIB SHAFIQUE • *State Key Laboratory of Crop Gene Resources and Breeding/National Key Facility for Crop Gene Resources and Genetic Improvement, Institute of Crop Sciences, Chinese Academy of Agricultural Sciences, Beijing, China*

MOTOKI SHIMIZU • *Iwate Biotechnology Research Center, Kitakami, Japan*

KEN SHIRASU • *RIKEN Center for Sustainable Resource Science (CSRS), RIKEN-TRIP, Yokohama, Kanagawa, Japan; Graduate School of Science, The University of Tokyo, Tokyo, Japan*

LIN-JIE SHU • *Department of Biology, University of Fribourg, Fribourg, Switzerland*

YING TANG • *National Key Laboratory of Agricultural Microbiology, Hubei Hongshan Laboratory, Hubei Key Laboratory of Plant Pathology, College of Plant Science and Technology, Huazhong Agricultural University, Wuhan, China*

RYOHEI TERAUCHI • *Laboratory of Crop Evolution, Graduate School of Agriculture, Kyoto University, Kyoto, Japan; Iwate Biotechnology Research Center, Kitakami, Japan*

KENICHI TSUDA • *National Key Laboratory of Agricultural Microbiology, Hubei Hongshan Laboratory, Hubei Key Laboratory of Plant Pathology, College of Plant Science and Technology, Huazhong Agricultural University, Wuhan, China*

ZHIYAN WEN • *State Key Laboratory of Plant Environmental Resilience, College of Biological Sciences, China Agricultural University, Beijing, China*

YUJING WU • *Department of Agri-microbiomics and Biotechnology, State Key Laboratory of Microbial Diversity and Innovative Utilization, Institute of Microbiology, Chinese Academy of Sciences, Beijing, China; University of the Chinese Academy of Sciences, Beijing, China*

YIJIA YAN • *Basic Immune System of Plants, Max Planck Institute for Plant Breeding Research, Cologne, Germany*

SUJI YE • *Department of Molecular, Cellular, and Developmental Biology, University of Michigan, Ann Arbor, MI, USA*

AYAKA YOSHIHISA • *Graduate School of Agriculture, Kindai University, Nara, Japan*
HIROFUMI YOSHIOKA • *Graduate School of Bioagricultural Sciences, Nagoya University, Chikusa, Nagoya, Aichi, Japan*
KEIKO YOSHIOKA • *Department of Cell & Systems Biology, University of Toronto, Toronto, ON, Canada*
MIKI YOSHIOKA • *Laboratory of Plant Immunology, Graduate School of Bioagricultural Sciences, Nagoya University, Chikusa, Nagoya, Aichi, Japan*
FAN-YU YU • *Department of Biology, University of Fribourg, Fribourg, Switzerland*
DINGLIANG ZHANG • *State Key Laboratory of Plant Environmental Resilience, College of Biological Sciences, China Agricultural University, Beijing, China*
XINYU ZHANG • *State Key Laboratory of Plant Environmental Resilience, College of Biological Sciences, China Agricultural University, Beijing, China*
YONGLIANG ZHANG • *State Key Laboratory of Plant Environmental Resilience, College of Biological Sciences, China Agricultural University, Beijing, China*

Chapter 1

Lumi-Map, a Luciferase Bioluminescence Screen of Mutants Combined with MutMap Allows Rapid Dissection of Plant Immune Signaling Pathways

Hiroaki Kato, Motoki Shimizu, Akira Abe, and Ryohei Terauchi

Abstract

Genetic screening of mutants and identification of their causative genes are effective strategies for understanding plant signaling pathways. Here, we present Lumi-Map, a method that combines real-time luciferase bioluminescence monitoring with MutMap analysis to identify single nucleotide polymorphisms (SNPs) responsible for the mutant phenotypes. Real-time monitoring of reporter gene expression enables high-throughput screen of mutants with altered transcriptional responses. Once candidate mutants are isolated, a whole-genome sequencing-based method MutMap allows for rapid identification of the causative SNPs. Using reporter lines driven by defense-related gene promoters, Lumi-Map successfully identified genes involved in pattern-triggered immunity, including a novel receptor gene for pathogen-associated molecular patterns. By selecting appropriate luciferase reporter plants, Lumi-Map is expected to serve as a versatile platform for dissecting diverse signal transduction pathways in plants.

Key words Lumi-Map, Genetic screening, Luciferase bioluminescence monitoring, Whole-genome sequencing, Signal transduction pathways

1 Introduction

Genetics represents a powerful and indispensable tool for elucidating genes involved in diverse plant biological processes. Forward genetic approaches have been extensively employed to identify components of signal transduction pathways [1]. Typically, target mutants are isolated from mutagenized populations. Given that large-scale mutant screening requires efficient phenotyping, methods that enable rapid and straightforward visual evaluation of morphological changes have proven particularly effective [2–4]. Mutant screening has also been conducted by assessing phenotypic alterations in response to biotic stress (e.g., pathogen inoculation), chemical treatments, or environmental changes [5–7]. In addition to visual phenotyping, non-visual approaches have also

Tsutomu Kawasaki and Hirofumi Yoshioka (eds.), *Plant Immunity: Methods and Protocols*, Methods in Molecular Biology, vol. 3012,
https://doi.org/10.1007/978-1-0716-5138-4_1,

been utilized to isolate mutants. However, analysis of non-visible traits, such as gene expression levels or metabolite accumulation, generally requires labor-intensive extraction and detection protocols, rendering such methods impractical for high-throughput screening. To facilitate convenient analysis of phenotype, stable transgenic plants expressing reporter constructs have been developed and widely employed in genetic screens. These reporter-based systems have enabled the identification of key regulatory genes involved in the circadian rhythm [8], hormone signaling [9], and stress response pathways. For instance, Ranf et al. (2012) utilized *Arabidopsis thaliana* lines expressing the calcium-sensitive reporter gene aequorin to monitor cytosolic Ca^{2+} dynamics. Through this approach, 35 mutant lines exhibiting altered calcium responses upon treatment with flg22, a peptide epitope derived from bacterial flagellin, were isolated [10, 11]. Furthermore, Li et al. (2014) established a screening platform named *Aggie* (*A*rabidopsis *g*enes *g*overning *i*mmune gene *e*xpression), in which the promoter of *FLG22-INDUCED RECEPTOR-LIKE KINASE 1* was fused to the firefly *luciferase* (*LUC*) gene. flg22-induced expression was subsequently quantified using a luminometer [12, 13]. These forward genetics approaches were coupled with traditional map-based cloning techniques and whole-genome sequencing for gene identification.

The next step following mutant screening is the identification of the causative gene. With the advancement of next-generation sequencing (NGS) technologies, the application of methods to isolate causative genes from mutants has expanded. Among these, MutMap has emerged as a rapid and powerful approach [14, 15]. In the MutMap method, mutant plants are crossed with the original wild-type plant and then self-pollinated. NGS is subsequently performed on the F_2 individuals exhibiting the mutant phenotype to identify the causative mutation.

Recently, we established Lumi-Map, a platform that combines a luciferase-based mutant screening system with gene identification by MutMap (Fig. 1), and successfully identified genes involved in the defense response triggered by flg22 treatment [16]. Using Lumi-Map, we simultaneously identified the apoplastic ceramidase NEUTRAL CERAMIDASE 2 (NCER2) and the plasma membrane receptor kinase RESISTANT TO DFPM-INHIBITION OF ABSCISIC ACID SIGNALING 2 (RDA2) [17], and elucidated the mechanism by which plants recognize the microbe-specific sphingolipid Pi-Cer D [18, 19]. These findings demonstrate the potential of Lumi-Map to significantly accelerate research in plant stress response mechanisms. By altering the combination of stress treatments and corresponding luciferase reporter lines, Lumi-Map is expected to contribute broadly to the elucidation of various plant physiological processes.

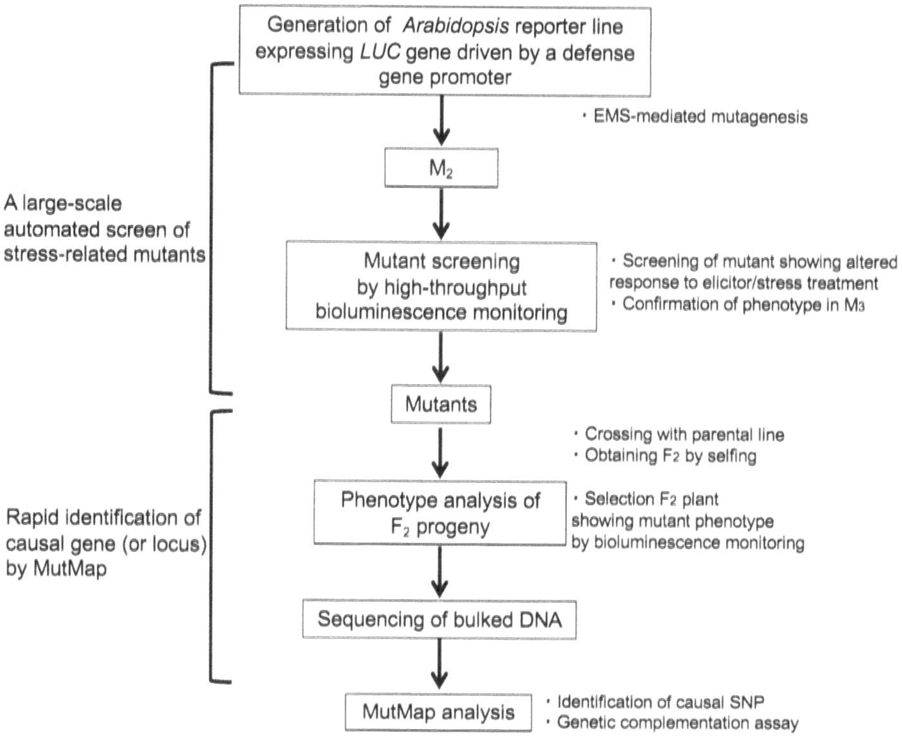

Fig. 1 Outline of the Lumi-Map method. An *Arabidopsis* luciferase reporter line was mutagenized using ethylmethansulfonate (EMS) to generate M_2 progeny. Mutant screening was conducted by monitoring bioluminescence in stress-treated M_2 seedlings. M_2 seedlings displaying altered bioluminescence phenotypes were propagated to the M_3 generation, and mutants were selected after phenotype confirmation. Causal gene identification was performed using MutMap. The selected mutants were crossed with the parental reporter line, and the resulting F_1 progeny were self-pollinated to obtain F_2 seeds. F_2 seedlings were subjected to stress treatment, and their bioluminescence phenotypes were monitored. Genomic DNA from F_2 individuals exhibiting mutant phenotypes was extracted and analyzed by whole-genome sequencing, followed by MutMap analysis to identify causal SNPs

2 Materials

2.1 Phenotyping Analysis by Bioluminescence Monitoring

1. *Arabidopsis thaliana* seeds of luciferase reporter lines (Col-0 background) (*see* **Notes 1** and **2**).
2. Bleach solution: 0.3% sodium hypochlorite solution supplemented with 0.02% Triton X-100.
3. Sterile water.
4. 96-well microplate (Luminunc™ Plates White F96; Thermo Fisher Scientific).
5. Murashige and Skoog (MS) liquid medium: half-strength liquid MS medium containing 0.5% sucrose.
6. 5 mM D-luciferin-K solution (Biosynth).
7. Parafilm.

8. Growth chamber.
9. Equipment of bioluminescence monitoring system (model CL96-4 and CI-08L, Churitsu Electric Corp, *see* **Note 3**).
10. Elicitor chemicals: flg22 (synthetic peptides, Life Technology Japan), chitin (LMC, Yaizu Suisankagaku Industry), Pi-Cer D (purified from *Phytophthora infestans*) (*see* **Note 4**).
11. Plate seals (Excel Scientific).
12. Ethylmethanesulfonate (EMS).
13. 0.1% Agar.
14. Soil or rockwool (Grodan; http://www.grodan.com).

2.2 MutMap Analysis

1. DNA extraction kit: e.g., DNeasy Plant Mini Kit (QIAGEN).
2. Library preparation kit for paired-end (PE) short-read sequencing: kits from Illumina or New England Biolabs (*see* **Note 5**).
3. Next-generation sequencing platform: e.g., NovaSeq X system (Illumina) (*see* **Note 5**).
4. MutMap pipeline: https://github.com/YuSugihara/MutMap.

3 Methods

3.1 Setting Condition for Bioluminescence Measurement

1. Sterilize seeds of *Arabidopsis* luciferase reporter lines by immersing them in a bleach solution and gently agitating for 10 min at room temperature.
2. Discard the bleach solution and rinse the seeds several times with sterile distilled water.
3. Incubate surface-sterilized seeds at 4 °C in the dark for 2–4 days.
4. Prepare MS liquid medium supplemented with D-luciferin-K to a final concentration of 50 μM.
5. Sow the seeds into wells of a 96-well microplate, each containing 150 μL of the prepared MS liquid medium.
6. Seal the microplate and its lid with parafilm to prevent evaporation and contamination.
7. Incubate the sealed microplate in a growth chamber at 23 °C under continuous light for 7 days.
8. Apply the elicitor chemical to the seedlings as appropriate, then seal the plate with a plate seal (*see* **Note 6**).
9. Measure bioluminescence using a bioluminescence monitoring system (*see* **Notes 7**, **8** and **9**).

3.2 Mutagenesis of Arabidopsis Luciferase Bioluminescence Reporter Line

1. Soak seeds of the luciferase bioluminescence reporter line in 0.05% Triton X-100 solution.
2. Rinse the seeds thoroughly with sterile distilled water several times to remove Triton X-100.
3. Treat the seeds with 0.15–0.3% (v/v) ethylmethanesulfonate (EMS) and gently mix by inverting for 15 h at 23 °C (*see* **Note 10**).
4. Discard EMS solution and rinse the seeds several times with water.
5. Incubate the EMS-treated seeds (M_1 seeds) at 4 °C in the dark for 2–4 days (*see* **Note 11**).
6. Suspend the M_1 seeds in 0.1% agar.
7. Sow the M_1 seeds evenly onto soil (*see* **Notes 12** and **13**).
8. Grow M_1 plants under standard conditions.
9. Harvest M_2 seed pools (*see* **Note 14**) (Fig. 2).

3.3 Screening of Mutants

1. Conduct phenotypic analysis of M_2 seedlings to screen for mutant candidates with altered bioluminescence response following treatment with elicitor chemicals or abiotic/biotic stress (*see* **Note 15**).
2. Transplant M_2 seedlings showing altered phenotypes (putative mutants) to soil and grow them to maturity for M_3 seed collection.
3. Perform phenotypic reconfirmation of the mutant candidates using M_3 seedlings.

3.4 MutMap analysis

1. Cross the selected mutant with the parental reporter line.
2. Harvest F_1 seeds.
3. Grow F_1 plant in soil (*see* **Note 16**).
4. Harvest F_2 seeds and select seeds of uniform size and shape using filter paper as a working surface (*see* **Note 17**).
5. Surface-sterilize the F_2 seeds along with seeds of the parental reporter line.
6. Grow the parental reporter line on soil for several weeks to obtain control leaf samples.
7. Sow F_2 seeds into MS liquid medium supplemented with D-luciferin and incubate for 7 days.
8. Evaluate the bioluminescence phenotype of F_2 individuals after treatment with the elicitor chemical using a bioluminescence monitoring system.
9. Select approximately 30 F_2 individuals showing the mutant phenotype (*see* **Note 18**).

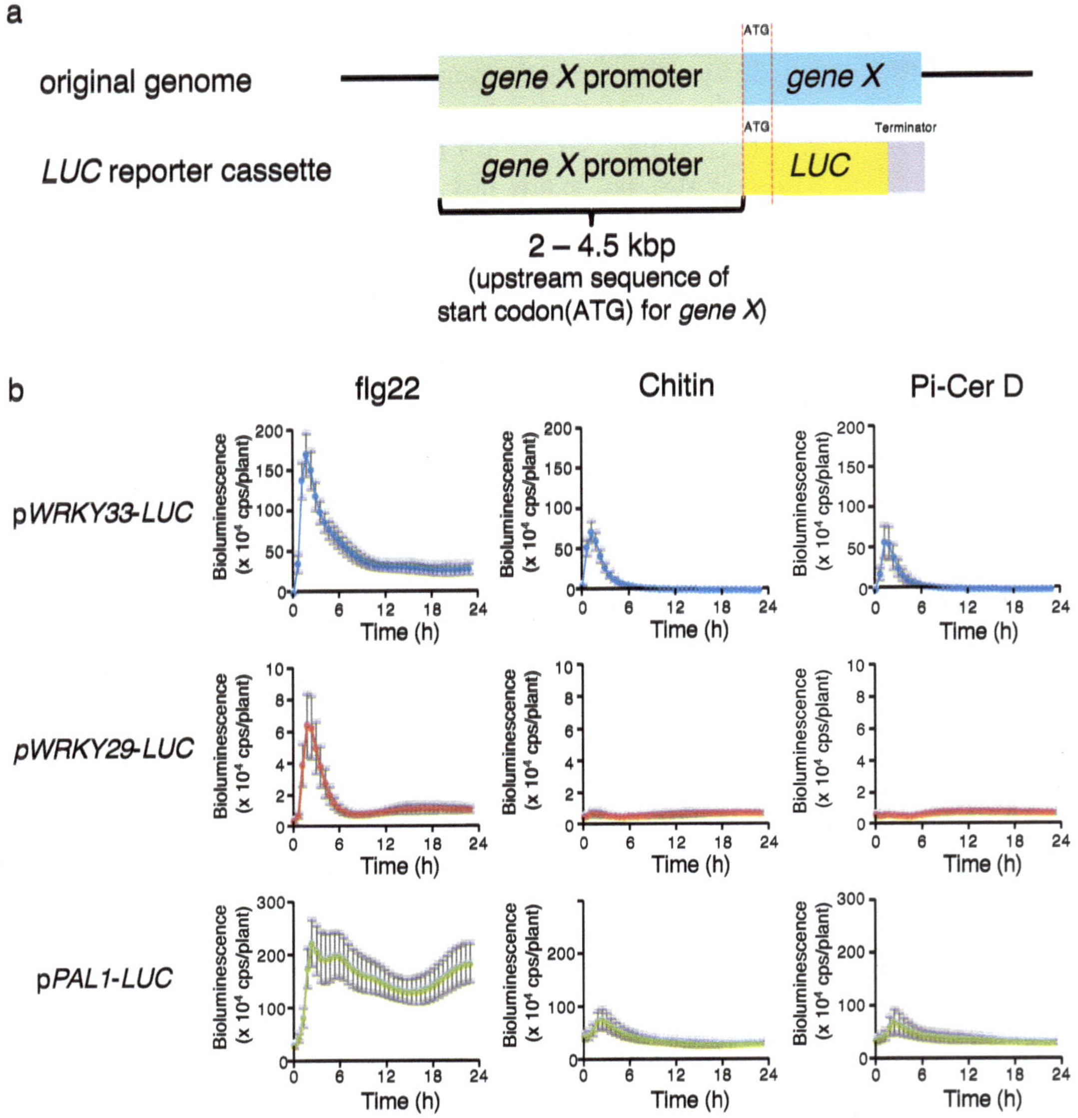

Fig. 2 *Arabidopsis luciferase* reporter lines. (**a**) Gene structure of the *luciferase* (*LUC*) reporter cassette. (**b**) Bioluminescence patterns of the p*WRKY33-LUC*, p*WRKY29-LUC*, and p*PAL1-LUC* reporter lines. Eight-day-old seedlings were treated with 100 nM flg22, 20 μg/mL chitin, or 0.17 μM Pi-Cer D. Bioluminescence from each seedling was monitored using a real-time bioluminescence monitoring system at the indicated time points

10. Grow the selected F_2 individuals in soil for several weeks.
11. Collect equal amount of leaf tissue from each of 30 selected F_2 individuals (*see* **Notes 19** and **20**).
12. Collect leaf tissue from the parental reporter line to serve as a reference genome in MutMap analysis.
13. Extract genomic DNA from both the bulked F_2 leaf samples and the parental reporter line using a DNA extraction kit.

14. Construct sequencing libraries for PE short-reads using commercially available library preparation kits according to the manufacturers' instructions (*see* **Note 5**).
15. Sequence the libraries using a short-read high-throughput sequencing system such as an Illumina NovaSeq X system (*see* **Note 21**).
16. Install and set up the MutMap analysis pipeline (*see* **Note 22**).
17. Perform MutMap analysis using the established MutMap pipeline (*see* **Note 23**).

3.5 Identification of Causative Mutation from MutMap Results

1. Select candidate SNPs located within the genomic region showing a peak in the SNP-index plot (Fig. 3), based on the following criteria: (a) SNPs predicted to have a strong functional impact (e.g., nonsense mutations > missense mutations > mutations in promoter or UTR regions); (b) SNPs located in close proximity to the peak of the SNP-index plot; (c) SNPs with an SNP-index value close to 1.
2. Generate complementation lines by introducing the wild-type version of the candidate gene into the mutant background, and evaluate restoration of the bioluminescence phenotype upon stress treatment (*see* **Note 24**).
3. Obtain T-DNA insertion lines for the candidate gene from a biological stock center (e.g., the Arabidopsis Biological Resource Center) and confirm whether these lines exhibit phenotypes similar to the mutant under stress conditions (*see* **Note 25**).

4 Notes

1. A reporter line that exhibits a stable and reproducible bioluminescence phenotype suitable for genetic analysis should be carefully selected and used.
2. Promoter regions typically consist of a 2–4.5 kb fragment upstream of the gene's start codon (ATG) (*see* Fig. 2a). In some cases, essential cis-regulatory elements for transcriptional activation may not be located upstream of the open reading frame (ORF) [20], resulting in undetectable bioluminescence. Thus, generating multiple reporter lines in parallel is recommended (*see* Fig. 2b).
3. A conventional 96-well plate reader capable of detecting chemiluminescence can also be used for bioluminescence measurements.
4. Depending on the screening objective, the elicitor chemicals or stress treatments may need to be altered accordingly.

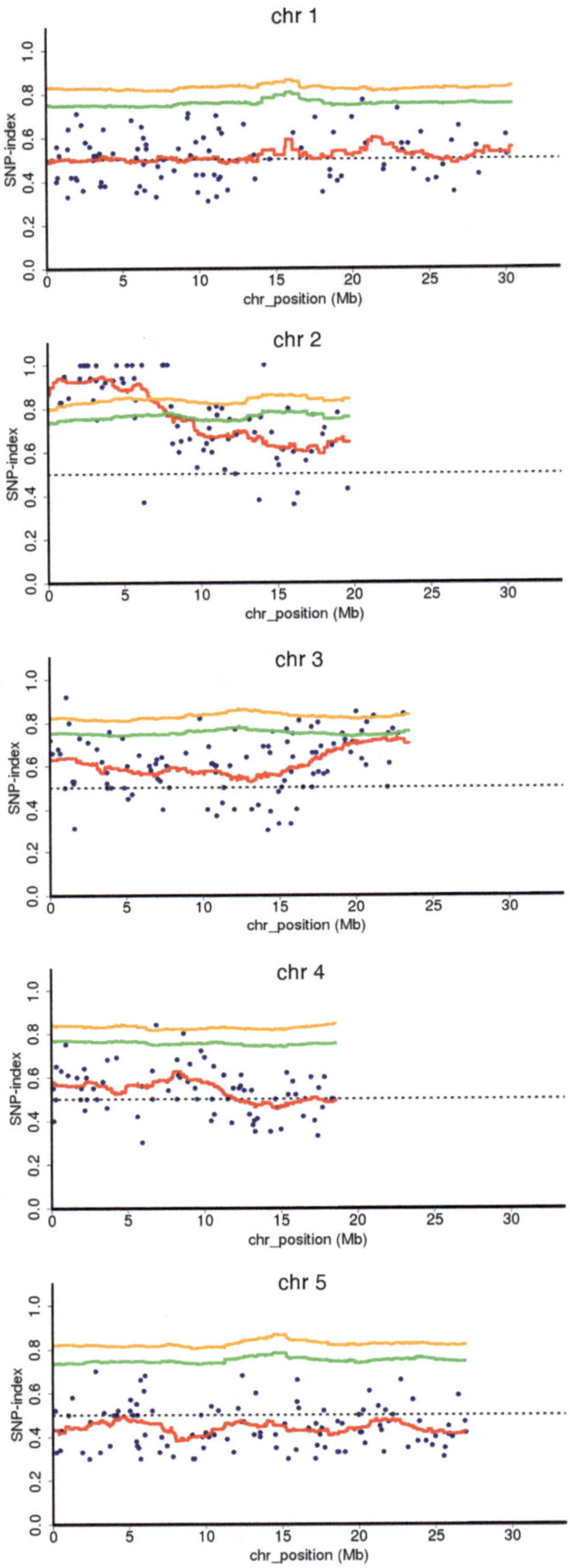

Fig. 3 Example of a typical MutMap SNP-index plot. SNP-index plots for the five *Arabidopsis* chromosomes generated by MutMap analysis, highlighting the genomic region with the highest SNP-index peak harboring

5. Construction of NGS libraries and NGS runs can be outsourced to commercial service providers.
6. For elicitor treatment, prepare chemical stock solutions at 10x the final concentration. Add one-tenth volume of the medium (e.g., 15 μL of chemical stock per 150 μL of medium).
7. Bioluminescence measurements should be recorded at 30–45 min intervals. Between measurements, incubate the culture plate under normal growth conditions (e.g., continuous light).
8. Stress treatments and culture conditions should be optimized according to the specific biological response being analyzed.
9. Long-term cultivation in 96-well microplates may impose stress on the plants, rendering them unsuitable for subsequent assays.
10. EMS treatment conditions directly affect mutation frequency. Excessive mutagenesis can introduce numerous background mutations, complicating MutMap analysis. We typically apply EMS treatment at multiple concentrations (e.g., 0.15% and 0.2%).
11. EMS-treated seeds are designated as M_1 seeds.
12. M_1 seeds are sown at a density of 10 seeds per 10 cm^2.
13. Rockwool can be used as an alternative to soil for plant cultivation.
14. M_2 seeds are collected in pools, with each pool consisting of seeds from approximately 200 M_1 plants. Assuming recessive mutations, screening 1000 plants from an M_2 pool derived from 200 M_1 plants theoretically covers all mutations.
15. In bioluminescence screening, mutants are isolated based on the maximum bioluminescence intensity at peak time and changes in the bioluminescence patterns.
16. High-quality seeds are essential for reliable bioluminescence phenotyping of F_2 progeny. Therefore, healthy F_1 plants should be grown to obtain good-quality F_2 seeds.
17. Consistency in seed quality is crucial for accurate phenotype analysis of seedlings.
18. For MutMap analysis, a bulk of 20–30 F_2 individuals is recommended to ensure sufficient recombination points. Including

Fig. 3 (continued) the candidate mutation. Blue dots represent SNPs identified in mutant lines relative to the wild-type reporter line. The red line indicates the average SNP-index calculated using a 2-Mb sliding window with 10-kb increments. The green and yellow lines represent the 95% and 99% confidence limits, respectively, under the null hypothesis of SNP-index = 0.5

individuals with incorrect phenotype in the bulk may reduce the accuracy of MutMap results.

19. To obtain healthy tissue for DNA extraction, newly emerged leaves are harvested from transplanted F_2 individuals showing the mutant phenotype.
20. Transplanted F_2 plants are grown to maturity to harvest F_3 seeds. F_3 plants showing the mutant phenotype are suitable for complementation and subsequent functional analysis, as they reduce background mutations unrelated to the phenotype. When phenotypic differences from wild type are subtle, performing MutMap with precise phenotyping using F_3 seedlings is effective.
21. Sufficient sequencing depth is necessary; a depth of approximately 30x generally yields reliable MutMap results.
22. The MutMap pipeline can be installed via Bioconda, which includes all required dependencies [15].
23. MutMap analysis requires the reference genome sequence of the parental lines. The reporter line reference genome sequence was constructed by substituting SNPs in the Col-0 reference with those found in the reporter line [16]. The Arabidopsis Col-0 reference genome was obtained from ftp://ftp.ensemblgenomes.org/pub/release-36/plants/fasta/Arabidopsis _thaliana/dna/Arabidopsis_thaliana.TAIR10.dna.toplevel.fa.
24. For complementation assays in reporter lines, use transformation vectors conferring antibiotic resistance different from that of the parental line. Complementation constructs should include genomic fragments containing candidate gene and its promoter region.
25. T-DNA insertion lines for candidate genes can be obtained from biological stock centers. Reporter gene expression in T-DNA insertion lines upon stress treatment should be analyzed by qRT-PCR.

Acknowledgments

We thank Prof. Kazuhito Kawakita and Makoto Ojika (Nagoya University, Japan) for Pi-Cer D purification. Computational analysis was partially conducted using the supercomputer at the National Institute of Genetics (NIG), Research Organization of Information and Systems (ROIS), Tokyo, Japan. This work was supported by the Precursory Research for Embryonic Science and Technology (PRESTO) program, grant number JPMJPR22D2 to H.K.

References

1. Page DR, Grossniklaus U (2002) The art and design of genetic screens: Arabidopsis thaliana. Nat Rev Genet 3:124–136
2. Saijo Y, Tintor N, Lu X, Rauf P, Pajerowska-Mukhtar K, Häweker H et al (2009) Receptor quality control in the endoplasmic reticulum for plant innate immunity. EMBO J 28:3439–3449
3. Dinh TT, Luscher E, Li S, Liu X, Won SY, Chen X (2014) Genetic screens for froral mutants in Arabidopsis thaliana: enhancers and suppressors. Methods Mol Biol 1110: 127–156
4. Fujita M, Tanabata T, Urano K, Kikuchi S, Shinozaki K (2018) RIPPS: a plant phenotyping system for quantitative evaluatiob of growth under controlled environmental stress conditions plant cell physiol 59:2030–2038
5. Glazebook J, Rogers EE, Ausubel FM (1996) Isolation of Arabidopsis mutants with enhanced disease susceptibility by direct screening. Genetics 143:973–982
6. Schreiber KJ, Austin RS, Gong Y, Zhang J, Fung P, Wang PW et al (2012) Forward chemical genetic screens in Arabidopsis identify genes that influence sensitivity to the phytotoxic compound sulfamethoxazole. BMC Plant Biol 12:226
7. Burke JJ, O'Mahony PJ, Oliver MJ (2000) Isolation of Arabidopsis mutants lacking components of acquired thermotolerance. Plant Physiol 123:575–587
8. Onai K, Okamoto K, Nishimoto H, Morioka C, Hirano M, Kami-Ike N et al (2004) Large-scale screening of *Arabidopsis* circadian clock mutants by a high-throughput real-time bioluminescence monitoring system. Plant J 40:1–11
9. Meier C, Bouquin T, Nielsen ME, Raventos D, Mattsson O, Rocher A et al (2001) Gibberellin response mutants identified by luciferase imaging. Plant J 25:509–519
10. Boller T, Felix G (2009) A renaissance of elicitors: perception of microbe-associated molecular patterns and danger signals by pattern-recognition receptors. Annu Rev Plant Biol 60:379–406
11. Ranf S, Grimmer J, Pöschl Y, Pecher P, Chinchilla D, Scheel D et al (2012) Defense-related calcium signaling mutants uncovered via a quantitative high-throughput screen in *Arabidopsis thaliana*. Mol Plant 5:115–130
12. Li F, Cheng C, Cui F, de Oliveira MV, Yu X, Meng X et al (2014) Modulation of RNA polymerase II phosphorylation downstream of pathogen perception orchestrates plant immunity. Cell Host Microbe 16:748–758
13. Feng B, Liu C, de Oliveira MV, Intorne AC, Li B, Babilonia K et al (2015) Protein poly (ADP-ribosyl)ation regulates arabidopsis immune gene expression and defense responses. PLoS Genet 11:e1004936
14. Abe A, Kosugi S, Yoshida K, Natsume S, Takagi H, Kanzaki H et al (2012) Genome sequencing reveals agronomically important loci in rice using MutMap. Nat Biotechnol 30:174–178
15. Sugihara Y, Young L, Yaegashi H, Natsume S, Shea DJ, Takagi H et al (2022) High-performance pipeline for MutMap and QTL-seq. PeerJ 10:e13170
16. Kato H, Onai K, Abe A, Shimizu M, Takagi H, Tateda C et al (2020) Lumi-Map, a real-time luciferase bioluminescence screen of mutants combined with MutMap, reveals *Arabidopsis* genes involved in PAMP-triggered immunity. Mol Plant Microbe Interact 33:1366–1380
17. Park J, Kim T-H, Takahashi Y, Schwab R, Dressano K, Stephan AB et al (2019) Chemical genetic identification of a lectin receptor kinase that transduces immune responses and interferes with abscisic acid signaling. Plant J 98: 492–510
18. Monjil MS, Kato H, Ota S, Matsuda K, Suzuki N, Tenhiro S et al (2024) Two structurally different oomycete lipophilic microbe-associated molecular patterns induce distinctive plant immune responses. Plant Physiol 196: 479–494
19. Kato H, Nemoto K, Shimizu M, Abe A, Asai S, Ishihama N et al (2022) Recognition of pathogen-derived sphingolipids in *Arabidopsis*. Science 376:857–860
20. Zhu Y, Klasfeld S, Jeong CW, Jin R, Goto K, Yamaguchi N et al (2020) TERMINAL FLOWER1-FD complex target genes and competition with FLOWERING LOCUS T. Nat Commun 11:5118

Chapter 2

Identification of Phosphorylation and Protein Interactions of Plant Pattern Recognition Receptors Through Immunoprecipitation

Yasuhiro Kadota and Ken Shirasu

Abstract

Plasma membrane-localized pattern-recognition receptors (PRRs) play a crucial role in detecting pathogen-associated molecular patterns (PAMPs). Among them, leucine-rich repeat (LRR) receptor-like kinases (RLKs) are particularly important in plants. However, their structural complexity and membrane association pose significant challenges for biochemical analysis. Many LRR-RLKs exhibit low expression levels in planta, and their hydrophobic surfaces contribute to instability, often requiring detergent-based extraction. These properties complicate co-immunoprecipitation (Co-IP), hindering the identification of interacting proteins and post-translational modifications. Recent advancements in immunoprecipitation technologies have improved the isolation of membrane proteins, enabling the analysis of protein-protein interactions and phosphorylation dynamics. Here, we describe an optimized immunoprecipitation protocol for studying plant plasma membrane receptor-like kinases, facilitating robust characterization of their interactomes and regulatory modifications.

Key words Immunoprecipitation, Protein phosphorylation, Receptor-like kinase

1 Introduction

The plasma membrane serves as a dynamic interface between plant cells and their external environment, enabling the perception of diverse environmental cues. To facilitate this, plants employ a vast array of plasma membrane-localized receptors. Among these, receptor-like kinases (RLKs) are transmembrane proteins composed of an N-terminal extracellular domain for ligand perception and a C-terminal intracellular kinase domain for signal transduction [1]. In contrast, receptor-like proteins (RLPs) share a similar architecture but lack a kinase domain [2]. These receptors regulate key physiological processes, including immune responses, hormone signaling, danger perception, and development [3, 4]. RLKs and RLPs primarily mediate signal transduction through

Tsutomu Kawasaki and Hirofumi Yoshioka (eds.), *Plant Immunity: Methods and Protocols*, Methods in Molecular Biology, vol. 3012,
https://doi.org/10.1007/978-1-0716-5138-4_2,

phosphorylation cascades, often involving other kinases as downstream targets [5–7].

Understanding the signal transduction downstream of membrane-localized receptors and their regulatory mechanisms requires the analysis of their protein complexes. However, isolating these complexes remains technically challenging due to their hydrophobic nature and the transient or weak interactions characteristic of kinases. Recent advances in co-immunoprecipitation (Co-IP) techniques have significantly improved the ability to capture protein interactors of RLKs and RLPs [8–13] and enabled the detection of key phosphorylation sites associated with receptor activation [14, 15].

Among these receptors, plant pattern recognition receptors (PRRs) detect pathogen-associated molecular patterns (PAMPs) at the plasma membrane. Notably, the leucine-rich repeat (LRR)-type RLKs FLAGELLIN-SENSING 2 (FLS2) and ELONGATION FACTOR-Tu RECEPTOR (EFR) recognize bacterial flagellin and elongation factor Tu (EF-Tu), respectively [16–18]. Perception of the immunogenic peptides flg22 (from flagellin) or elf18 (from EF-Tu) by FLS2 or EFR, respectively, triggers their rapid association with the co-receptor LRR-RLK BAK1, leading to reciprocal phosphorylation and the initiation of downstream immune responses, including a reactive oxygen species (ROS) burst, transcriptional reprogramming, and ultimately, enhanced immunity [6, 9, 19–23]. Co-IP of EFR has successfully identified several BAK1 homologs, the NADPH oxidase RBOHD (which mediates PAMP-induced ROS production), and regulatory RLKs such as IOS1 and QSK1 [8, 9, 13, 24]. These discoveries revealed a novel activation mechanism, in which EFR-associated kinase BIK1 activates RBOHD [9], while IOS1 and QSK1 function as regulatory components [13, 24]. Furthermore, immunoprecipitation (IP) of EFR followed by selected reaction monitoring with liquid chromatography-tandem mass spectrometry (LC-MS/MS) has enabled the identification of phosphorylation sites essential for EFR activation [14, 15]. Collectively, these findings highlight IP as a powerful approach for elucidating the regulatory mechanisms of RLKs and RLPs, providing crucial insights into their signal transduction pathways at the molecular level.

Although IP is valuable for identifying PRR interactors and phosphorylation sites, its specificity can be compromised by contamination from non-specific proteins. Major contaminants include antibodies (Fig. 1) and plant-derived sticky proteins (Fig. 2), both of which can obscure true interactors and reduce the reliability of Co-IP results. In conventional IP using antibodies bound to protein A/G, elution with boiled SDS buffer often results in the release of heavy and light chains into the eluate, whereas acidic elution minimizes antibody contamination because IgG remains largely associated with protein A/G under low-pH conditions (Fig. 1). A

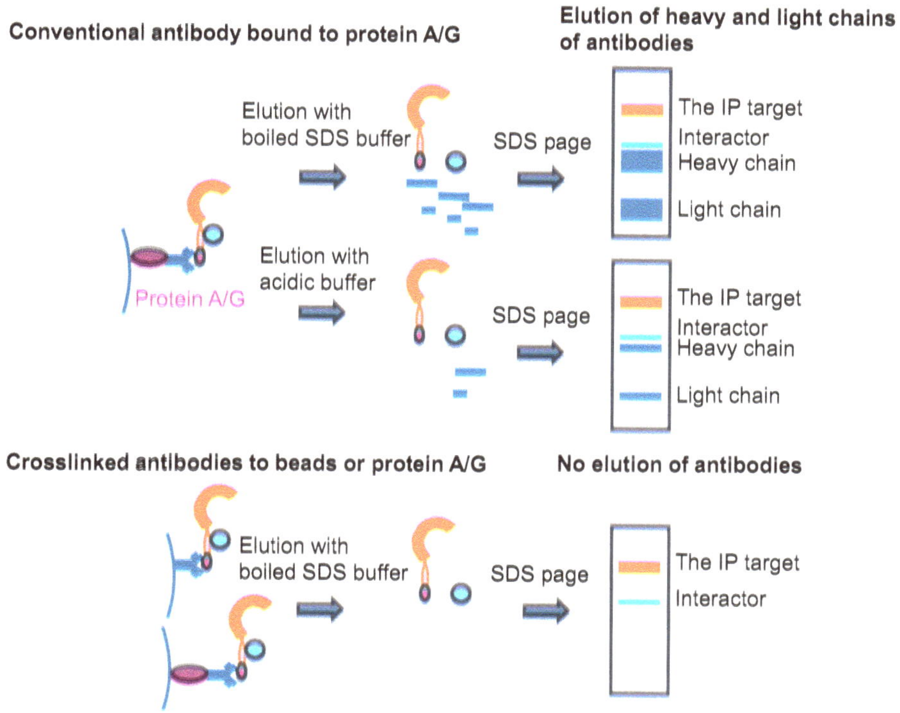

Fig. 1 Contamination risks of antibodies in conventional co-immunoprecipitation (Co-IP). Elution with boiled SDS buffer releases antibody heavy and light chains, which appear as contaminating bands in the eluate. Acidic elution minimizes antibody contamination because IgG remains largely bound to protein A/G under low-pH conditions. Crosslinking antibodies to beads prevents antibody release during boiled SDS elution

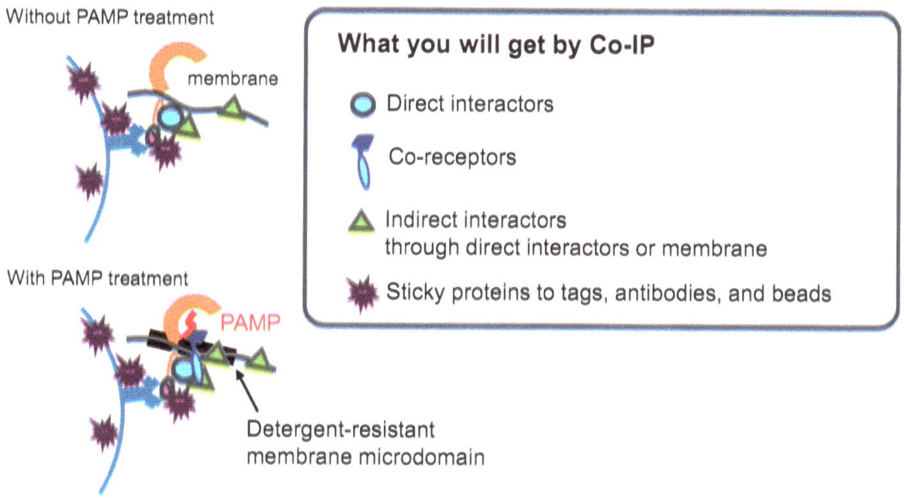

Fig. 2 Types of proteins isolated by Co-IP. Co-IP of RLKs can capture direct interactors, such as co-receptors, as well as indirect interactors that associate through direct interactors or membrane components. Additionally, nonspecific "sticky" proteins may bind to tags, antibodies, or beads, potentially affecting specificity

more effective strategy involves crosslinking antibodies to beads or protein A/G, which prevents release of antibodies during elution and significantly improves the specificity and clarity of interactome analysis.

In addition to antibody contaminants, plant-derived sticky proteins (Fig. 2) present a significant challenge, as they can non-specifically bind to beads, antibodies, or affinity tags fused to the target protein, generating high background noise and hindering accurate identification of true interactors and phosphorylation sites. Moreover, Co-IP captures not only direct interactors but also indirect interactors that associate through intermediary proteins or membrane compartments. For example, the pattern recognition receptor FLS2 localizes to a specific plasma membrane microdomain and, upon flg22 recognition, relocates to detergent-resistant membrane (DRM) microdomains [25]. Consistently, Co-IP using FLS2 or EFR has identified multiple proteins previously reported to accumulate in DRM fractions following flg22 treatment [13, 26]. These observations underscore the critical need to consider membrane organization and dynamics when interpreting IP results.

To improve the reliability of IP-based analyses, several strategies can be implemented to minimize contamination of sticky proteins (Fig. 3). First, magnetic beads offer a cleaner alternative

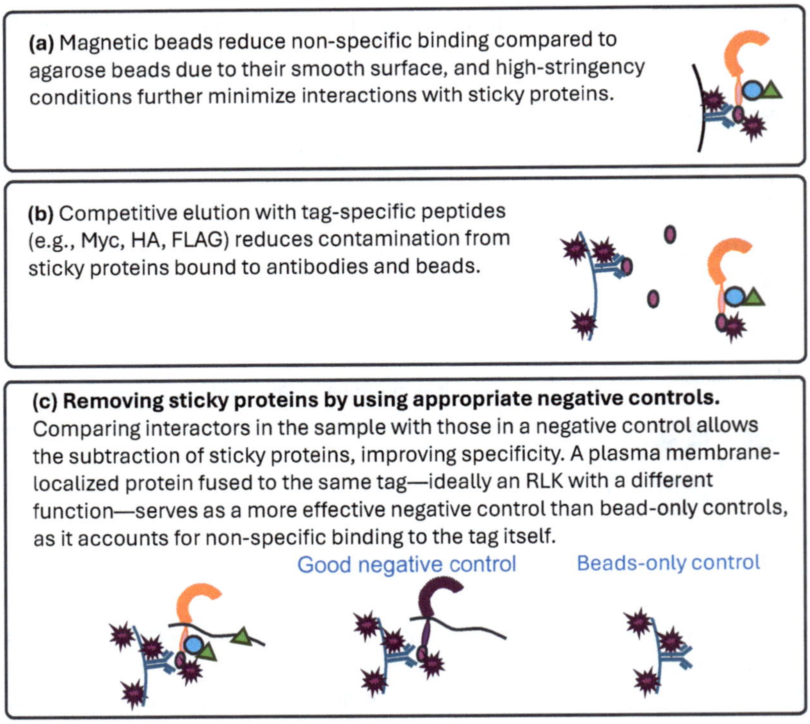

Fig. 3 Guidelines to reduce contamination of sticky proteins

to traditional agarose beads, as their smooth surfaces reduce non-specific protein adsorption (Fig. 3a). In addition, magnetic separation also allows gentler washing, helping preserve weak or transient interactions. Second, the use of high-stringency binding and washing buffers can further reduce interactions with sticky proteins. Careful optimization of salt concentrations, detergent composition, and pH is critical to balance stringency with the preservation of genuine interactions. Third, competitive elution using a synthetic peptide corresponding to the affinity tag, instead of harsh treatments such as boiling in SDS buffer, can effectively reduce the co-elution of sticky proteins bound to antibodies or beads (Fig. 3b). This method ensures the selective recovery of target protein complexes while minimizing background contamination. Lastly, bioinformatic subtraction of non-specific proteins by comparing experimental samples with appropriate negative controls is essential. The most effective negative controls are proteins with similar molecular weight, expression level, and subcellular localization as the protein of interest (Fig. 3c), allowing for accurate discrimination between true interactors and background binders.

The selection of an appropriate epitope tag and high-quality antibodies is critical for the success of IP (Figs. 4 and 5). We do not recommend using homemade antibodies for IP-based interactome studies unless their titer and specificity are comparable to those of commercial antibodies. A more reliable strategy involves expressing the tagged protein under the control of its native promoter in plants to preserve physiologically relevant interactions. Even among commercial antibodies, performance can vary significantly. Antibodies that exhibit minimal non-specific binding in immunoblotting are generally more suitable for IP, as they reduce the risk of co-purifying contaminant proteins. The choice of tags plays a crucial role, especially when validating interactions by immunoblotting. Some antibody-tag combinations, such as α-HA-HRP (clone 3F10 monoclonal antibody), exhibit clean signals with minimal background, whereas others may produce non-specific bands or high background noise. For instance, α-FLAG-HRP (clone M2 monoclonal antibody, A8592, Sigma-Aldrich) tends to detect non-specific proteins in both *Arabidopsis thaliana* and *Nicotiana benthamiana*, while α-GFP (Rabbit polyclonal antibody ab290, Abcam) often generates elevated background signals on PVDF membranes (Immun-Blot® PVDF, Bio-Rad) under our experimental conditions. If the target protein is highly abundant in the sample, such background signals may be negligible. However, for low-abundance proteins or those with molecular weights similar to non-specific bands, background signals can obscure detection and compromising the accuracy of data interpretation (Figs. 4 and 5). In addition, for the identification of PRR modifications, such as phosphorylation, maximizing protein yield while preventing

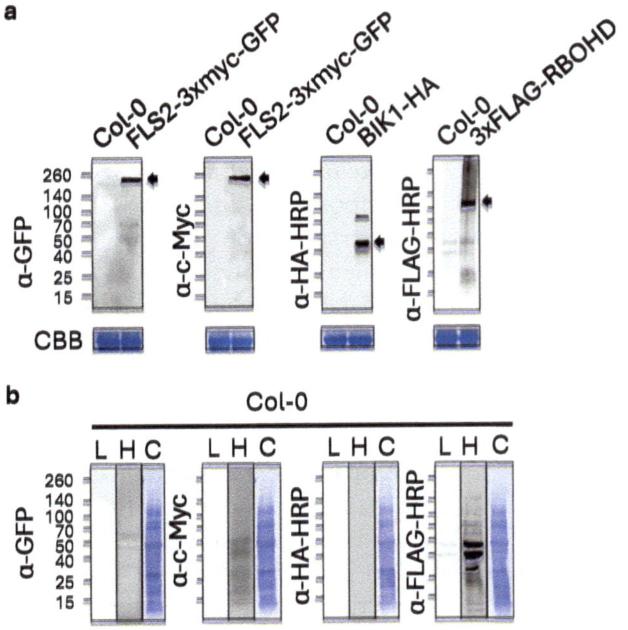

Fig. 4 Detection of tagged proteins and non-specific bands in *Arabidopsis thaliana* by immunoblotting using α-GFP, α-c-Myc, α-HA-HRP, and α-FLAG-HRP. (**a**) Immunoblot analysis of FLS2-3×Myc-GFP (*pFLS2:gFLS2-3×Myc-GFP*) [29], BIK1-HA (*pBIK1:gBIK1-HA*) [33], and 3×FLAG-gRBOHD (*rbohD/pRBOHD:3xFLAG-gRBOHD*) in seedlings of stable transgenic plants [9]. Protein levels were assessed using α-GFP (x5,000, ab290, Abcam, Cambridge, UK) with α-rabbit secondary antibodies (x10,000, NA934V Anti-Rabbit IgG, HRP-Linked Whole Ab Donkey, Cytiva), and α-c-Myc (x10,000, 9E10 SC-40, Santa Cruz Biotechnology) with α-mouse secondary antibodies (x10,000, Anti-Mouse IgG, HRP-Linked Whole Ab Sheep, Cytiva), as well as α-HA-HRP (x5,000, 3F10, Roche, Basel, Switzerland) and α-FLAG-HRP antibodies (x2,000, A8592, M2 monoclonal antibody, Sigma-Aldrich, St. Louis, MO, USA). Membranes were blocked with 5% skim milk in TBS-T and washed with TBS-T. Equal loading of protein samples is shown by coomassie brilliant blue (CBB) staining. Black arrows indicate expected protein bands. (**b**) Detection of non-specific bands in *Arabidopsis*. Non-specific bands were observed under lower (L) and higher (H) exposure conditions. Total protein was stained with CBB (C)

dephosphorylation is often more critical than purity, as SDS-PAGE can effectively separate contaminating proteins.

As discussed above, multiple factors—including IP systems, antibodies, experimental conditions, and appropriate negative controls—must be carefully selected and optimized to achieve reliable IP results (Fig. 6). We previously published protocols for the IP of plasma membrane-localized PRR complexes [27], but here, we present updated protocols for the IP of PRR complexes, utilizing different systems tailored to distinct experimental goals.

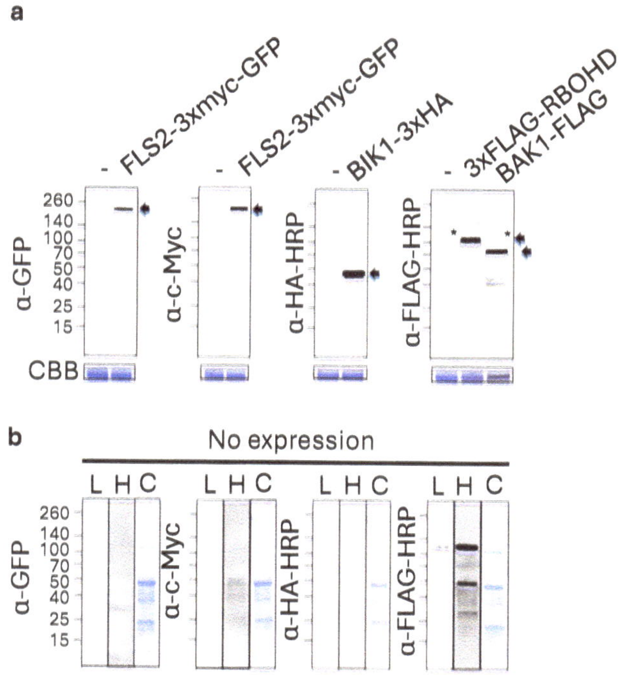

Fig. 5 Detection of tagged proteins and non-specific bands in *Nicotiana benthamiana* by immunoblotting using α-GFP, α-c-Myc, α-HA-HRP, and α-FLAG-HRP. (**a**) Immunoblot analysis of FLS2-3×Myc-GFP (*pCAMBIA-pFLS2::gFLS2-3xmyc-GFP*) [29], BIK1-3xHA (*pGWB14-35S-cBIK1-3xHA*), 3×FLAG-RBOHD (*pBin19g-pRBOHD::gRBOHD-3xFLAG*) [9], and BAK1-FLAG (*pGWB11-35S-cBAK1-FLAG*). Protein detection was performed using α-GFP with α-rabbit secondary antibodies, α-c-Myc with α-rabbit secondary antibodies, as well as α-HA-HRP and α-FLAG-HRP antibodies. The proteins were transiently expressed in *N. benthamiana* leaves via co-infiltration of *Agrobacterium* solution (OD = 0.6), mixed with p19-expressing *Agrobacterium* in a 1:1 ratio. Immunoblotting conditions were the same as in Fig. 4. Equal loading of protein samples is shown by coomassie brilliant blue (CBB) staining. Black arrows indicate expected protein bands, while asterisks (*) denote non-specific bands. (**b**) Detection of non-specific bands in *N. benthamiana*. Non-specific bands were observed under lower (L) and higher (H) exposure conditions. Total protein was stained with CBB (C)

2 Materials

2.1 Protein Extraction Components

1. Mortars and pestles.
2. Extraction buffer: 50 mM Tris–HCl pH 7.5, 100–150 mM NaCl, 10% glycerol, 5 mM DTT*, 1% (v/v) protease inhibitor cocktail (cOmplete™ ULTRA, Sigma-Aldrich)*, 0.5–2% (v/v) IGEPAL CA-630 (Sigma-Aldrich), 1 mM $Na_2MoO_4 \cdot 2H_2O$*, 1 mM NaF*, 1.5 mM activated sodium orthovanadate (Na_3VO_4)*, 1 mM EDTA (Optional), 1% (v/v) protein

•Identification of PRR Modifications
→ Express tagged PRR in plants and use an agarose bead system (e.g., GFP-Trap from ChromoTek) to efficiently isolate a sufficient amount of PRR for modification analysis.

•Identification of Novel Components in the PRR Complex
→ Express tagged PRR in plants and use a magnetic bead system (e.g., MACS technology from Miltenyi Biotec) to isolate interacting proteins with high specificity.

•Investigating PRR-BAK1 Interaction
→ Use home-made antibodies against PRR and/or BAK1, and immunoprecipitate endogenous PRR or BAK1 using the TrueBlot system (eBioscience) to minimize interference from heavy and light chains of IgG.

→ Alternatively, express tagged PRR in plants and immunoprecipitate PRR using either an agarose bead system or a magnetic bead system.

•Investigating PRR Interactions with Other Interactors
→ Express tagged PRR or interactors in plants and use a magnetic bead system (e.g., MACS technology from Miltenyi Biotec) for efficient and specific immunoprecipitation.

Fig. 6 Guidelines for immunoprecipitation strategies based on experimental goals

phosphatase inhibitor cocktail 2 and 3 from Sigma-Aldrich* (optional) (*see* **Note 1**).

N.B.: Do not add IGEPAL CA-630 to the extraction buffer if you homogenize samples with POLYTRON homogenizer to avoid making bubbles. Add it after homogenization (*see* **Note 2**).

Ingredients labeled with * should be added right before each use. Decide the concentration of NaCl depending on your purpose and the biochemical properties of your protein (*see* **Note 3**). Decide the best detergent and its concentration depending on your purpose (*see* **Note 4**).

3. Sand (50–70 mesh particle size, 274739 Sigma-Aldrich) (Optional).
4. POLYTRON Homogenizer (PT-DA 12/2EC-E157, KINEMATICA) (Optional).
5. Empty 20 mL column with 30 μm polyethylene filter (Econo column, 732-1010, BIO-RAD).
6. Phenylmethanesulfonyl fluoride (PMSF) 100 mM in isopropanol.
7. Rotator (RT-50, TAITEC).
8. Bradford Protein Assay kit (BIO-RAD).

2.2 Magnetic Beads Co-IP System from Miltenyi Biotec

1. Rotator (RT-50, TAITEC).
2. Low protein binding tubes 1.5 mL (72.706.600, SARSTEDT).
3. μMACS Separator (Miltenyi Biotec).
4. μ-columns (Miltenyi Biotec).

5. μMACS GFP beads (Miltenyi Biotec).
6. Empty 20 mL-column 30 μm polyethylene filter (Econo column, 732-1010, BIO-RAD).
7. Parafilm.
8. Elution buffer (SDS buffer from Miltenyi Biotec): 50 mM Tris–HCl (pH 6.8), 50 mM DTT, 1% SDS, 1 mM EDTA, 0.005% bromphenol blue, 10% glycerol.
9. Peristaltic pump with a small tube to connect to μ-columns (SJ-1211II, ATTO).

2.3 IP Using GFP-Trap Agarose Beads from ChromoTek

1. Rotator (RT-50, TAITEC).
2. Low protein binding tubes 1.5 mL (72.706.600, SARSTEDT).
3. GFP antibody conjugated to agarose beads (GFP-trap_A, Chromotek).
4. Mini Bio-Spin® Chromatography Columns (#7326207, BIO-RAD).
5. LDS Sample Buffer (NuPAGE®, Invitrogen).
6. Sample Reducing Agent (NuPAGE®, Invitrogen).

2.4 Co-IP with TrueBlot System from eBioscience

1. Rotator (RT-50, TAITEC).
2. Low protein binding tubes 1.5 mL (72.706.600, SARSTEDT).
3. True-blot anti-rabbit Ig beads (eBioscience).
4. TrueBlot Anti-IgG HRP Rabbit (Mouse, eBioscience).

3 Methods

3.1 Preparing Plant Material

1. Generate transgenic *Arabidopsis* plants expressing proteins fused to appropriate affinity tags to ensure efficient detection and immunoprecipitation (*see* **Notes 5** and **6**).
2. Sterilize the seeds.
3. Sow seeds on MS agar plates (1% sucrose) and keep them in the dark at 4 °C for 2–3 days.
4. Grow seedlings for 7 days under controlled conditions (120–130 μmol/m^2/s light intensity, 16 h light/8 h dark cycle, 22–23 °C).
5. Transfer seedlings into 6-well plates containing 5 mL MS medium with 1% sucrose per well (5 seedlings per well).
6. Grow for an additional 7–10 days (14–17 days total) before harvesting samples.

Prepare 5 plates per treatment (10–15 g fresh weight) for large-scale Co-IP or IP for phosphorylation site identification. Prepare 1–2 plates per treatment (2–3 g fresh weight) for small-scale Co-IP.

3.2 Ligand Treatment

1. Dilute PAMP stock solution in distilled water (1–10 μM for flg22 or elf18). If you aim to activate all receptors, using a higher concentration of PAMP would be more effective. One micromolar flg22 or elf18 is sufficient to observe BAK1 interaction, while more than 5–10 μM flg22 or elf18 is required for stomatal closure and PRR endocytosis [28–30].
2. Prepare two beakers of different sizes (e.g., 300 mL and 200 mL).
3. Place all *Arabidopsis* seedlings into the larger beaker after removing MS media with paper towels.
4. Add the PAMP solution or water (for mock treatment) to fully submerge the seedlings, then place the smaller beaker on top to ensure they remain underwater (*see* **Note 7**).
5. Place the beaker setup into a vacuum chamber and apply vacuum at −0.1 MPa for 2 min.
6. Slowly release the pressure and start timing. The PAMP solution will enter the plants, causing the leaves to darken. If infiltration is incomplete, repeat the vacuum infiltration process.
7. After 5 min (or 10–15 min, depending on the experiment), discard the PAMP solution.
8. Place the seedlings on paper towels and gently dry them.
9. Immediately freeze the seedlings in liquid nitrogen and store them at −80 °C.

   The same protocol for PAMP treatment and Co-IP can be applied to *N. benthamiana* leaves transiently expressing tagged proteins, using a different buffer (*see* **Notes 8** and **9**).

3.3 Grinding the Samples

1. Grind frozen tissues into a fine powder using a mortar and pestle with liquid nitrogen.
2. (Optional) Add 0.5 g polyvinylpolypyrrolidone (PVPP) to remove phenolic compounds. This step is particularly important for *N. benthamiana* samples, as their high phenolic content can interfere with protein extraction. However, it is generally unnecessary for *Arabidopsis* samples.
3. Transfer the powdered tissue into a 50 mL tube and make a small hole in the lid to prevent pressure buildup. The final powder volume should be approximately 35 mL.
4. Keep in −80 °C until protein extraction (*see* **Note 10**).

3.4 Protein Extraction

1. Prepare fresh extraction buffer (40 mL per sample for large-scale IP and 15 mL for small-scale IP).
2. Add extraction buffer (2 mL/g tissue) to the tube containing ground powder, then immediately add PMSF to a final concentration of 1 mM. Important: PMSF degrades within 30–60 min in aqueous solutions; fresh PMSF should be added immediately before extraction.
3. Quickly mix using a spoon and vortex to ensure thorough incorporation of the ground powder with the extraction buffer. Note: Powder that melts without mixing may degrade or undergo dephosphorylation. Replace the lid with a new one without a hole.
4. (Optional) Homogenize samples with POLYTRON homogenizer (*see* **Note 2**). If using this step, prepare extraction buffer without IGEPAL and add it after homogenization.
5. Incubate at 4 °C for 1 h with gentle rolling to solubilize the plasma membrane.

3.5 Centrifugation of Crude Extracts

1. Centrifuge at 15,000× *g*, 4 °C for 5 min and transfer the supernatant to a new tube. Repeat centrifugation to further clarify the extract.
2. Pass the extract through Miracloth (475855, Millipore) and repeat centrifugation.
3. Pass the extract through an empty column with 30 μm polyethylene filter (Econo-Pac® Chromatography Columns, Bio-Rad) to trap residual cell debris.
4. Continue centrifugation at 15,000× *g*, 4 °C for 5 min, repeating until no visible precipitate remains.

   Important: This step is critical, as any remaining precipitate cannot be removed later and would significantly increase false positives. Precipitate containing PVPP is very fragile and can easily break during transfer to new tubes for the next centrifugation step. Additionally, PVPP particles are small enough to pass through polyethylene filters.

3.6 Determination of Protein Concentration

1. Determine the protein concentration using 1/10-diluted samples (Bradford Protein Assay kit, Bio-Rad).
2. Dilute samples to 3–5 mg/mL protein using extraction buffer and transfer to 15 mL or 50 mL tubes. A higher extract concentration can sometimes lead to protein aggregation during incubation with antibodies. Additionally, excessive protein concentration may increase non-specific interactions with the tube, columns, and beads, leading to higher background noise.

3. Take a small aliquot to use as input, mix it with SDS or LDS sample buffer, freeze it with liquid nitrogen, and keep it in the freezer before use.

3.7 IP

Selecting the most suitable IP system is essential for specific experimental goals (Fig. 6). We recommend Miltenyi Biotec's magnetic bead system for identifying protein interactors in plants (*see* **Notes 11** and **12**) and GFP-Trap from ChromoTek for identifying phosphorylation sites in RLKs (*see* **Note 13**). The magnetic beads (50 nm in diameter) feature a smooth surface, enabling rapid capture of tagged proteins while minimizing background interactions. However, their binding capacity is limited by the column format, making GFP-Trap a better choice for applications requiring higher protein recovery, such as phosphorylation studies. We previously compared both systems for isolating EFR-BAK1 complexes. GFP-Trap recovered more GFP-tagged EFR, whereas the magnetic bead system retrieved more BAK1 [27]. This suggests that the magnetic bead system better preserves native protein-protein interactions at the plasma membrane, likely due to milder washing conditions. Since antibodies are covalently coupled to beads in both systems, antibody contamination is minimized in the elution. Moreover, GFP-Trap uses camelid-derived antibodies, which are not recognized by conventional HRP-fused secondary antibodies, thereby further minimizing background signals in immunoblotting.

To investigate the interaction between endogenous FLS2 and BAK1 using homemade antibodies, we recommend the TrueBlot system from eBioscience because it minimizes interference from heavy and light chains of immunoglobulins, allowing for clearer detection of co-immunoprecipitated proteins in immunoblotting [8].

3.7.1 IP Using Magnetic Beads from Miltenyi Biotec

1. Add 50 μL (for small-scale IP) or 250 μL (for large-scale IP) of magnetic beads to the crude extract and incubate on a rotator for 0.5–1 h at 4 °C.
2. Securely connect μ Columns and empty columns with a white filter (20 mL Econo-Pac® Chromatography Columns, Bio-Rad) using parafilm (Fig. 7). Note: The magnetic beads are extremely small and can pass through 30 μm polyethylene filter.
3. Place the assembled columns in the magnetic field of a μMACS Separator and bring them to a cold room.
4. Equilibrate the assembled columns by applying 1–5 mL of extraction buffer.
5. Apply the crude extract to the assembled column and discard the flow-through.

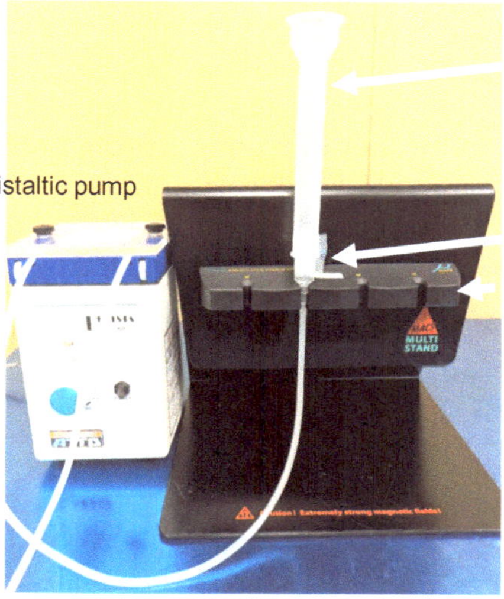

Fig. 7 Assembly of columns and peristaltic pump on the magnetic µMACS separator. A 20 mL empty column (Econo-Solomon, BIO-RAD) is tightly connected to a µ Column to allow the application of a large volume of crude extract. The magnetic beads are small enough to pass through the filter of the empty column, while cell debris is retained, even if some contamination occurs. If the flow rate is too slow or to ensure complete removal of liquid during washing, suction can be applied using a peristaltic pump from the bottom of the µ Column

Important: From this step onward, DO NOT remove the column from the µMACS Separator, as beads and immunoprecipitated proteins will pass through the column if separated from the magnetic field.

6. If the flow rate is very slow, apply suction from the bottom of the column using a peristaltic pump. Remove the empty column from the µ Column after the entire extract has passed through.
7. Wash the µ Column 3–5 times with 500 µL of extraction buffer (*see* **Note 11**).
8. If the outer surface of the µ Column or µMACS Separator gets wet with the extract, clean it using a pipette or paper towel to prevent contamination of the elution.

3.7.2 Elution from Magnetic Beads with SDS Buffer

1. Heat the elution buffer from Miltenyi Biotec on a heat block at 99 °C.
2. Remove any remaining buffer in the µ Column by applying suction with a peristaltic pump from the bottom of the µ Column.
3. Bring the columns and the µMACS Separator next to the heat block.

4. Place a low-protein binding tube under the μ Column.
5. Add 20 μL of heated elution buffer to the μ Column and immediately press the column slightly with your thumb or gloved fingers, ensuring that the boiled elution buffer reaches the black part of the μ Column where the magnetic beads are located.
6. Wait for 5 min, then press the column again to elute.
7. Add another 20–30 μL of heated elution buffer and repeat the process.
8. Divide the elution buffer containing proteins into multiple tubes, freeze them in liquid nitrogen, and store them in the freezer until use.

   Note: Freeze-thaw cycles can significantly reduce the signal for some proteins, so avoid repeated freezing and thawing to preserve protein integrity.

3.7.3 Determining the Total Amount of Beads Collected in the μ Column

1. Remove the column from the μMACS Separator and place it into a new 1.5 mL tube.
2. Add 1 mL of extraction buffer into the column and gently press with your fingers. The brown-colored beads will be released from the column.

3.7.4 Reusing μ Columns

1. Extensively wash μ Columns with detergent and water after you use them.
2. Ensure complete drying by applying 100% ethanol to remove all residual water and prevent rusting.

   Note: We do not recommend reusing μ Columns for discovering interactors, but they can be reused for IP experiments to validate interactions.

3.7.5 IP Using GFP-Trap Agarose Beads from ChromoTek

1. Remove storage buffer and wash agarose beads 2–3 times with extraction buffer.
2. Add 50 μL of 50% agarose to each 15 mL crude extract using a cut pipette tip and incubate on a rotator at 4 °C for 2–4 h.
3. Centrifuge 500× *g* for 1 min at 4 °C.
4. Carefully remove the supernatant using a pipette or vacuum pump.
5. Add a small volume of extraction buffer and transfer the agarose beads to a new 1.5 mL low-protein binding tube using a 1 mL cut pipette tip.
6. Rinse the 50 mL tube with 1 mL of fresh extraction buffer to recover any remaining agarose beads attached to the tube surface, then transfer to the 1.5 mL tube.

7. Centrifuge at 500× *g* for 30 s at 4 °C, remove the supernatant, and resuspend the beads in 1 mL of extraction buffer.
8. Wash the beads 2–4 more times with 1 mL of extraction buffer.
9. Remove the final wash buffer before proceeding to elution or further analysis. If complete removal of the washing buffer is required, use a 1 mL syringe with a 25G × 1 needle (0.5 × 25 mM) to carefully remove any remaining liquid.

3.7.6 Elution from GFP-Trap Agarose Beads

1. Prepare 2x LDS buffer with sample reducing agent.
2. Add 40 μL of 2x LDS buffer to the agarose beads.
3. Heat at 70 °C 15 min (*see* **Note 14**). For proteins highly attached to the plasma membrane, elution should be performed at lower temperatures than boiling.
4. Transfer the entire sample to a mini spin column and centrifuge for 1 min at maximum speed.
5. Add 10 μL of 2× LDS buffer to the agarose beads and centrifuge for 1 min at maximum speed to recover the remaining proteins.
6. Freeze the samples in liquid nitrogen and store them in the freezer until use.

Note: Freeze-thaw cycles can significantly reduce the signal for some proteins, so avoid repeated freezing and thawing to preserve protein integrity.

3.7.7 Co-IP Using TrueBlot System from eBioscience

1. Remove storage buffer and wash agarose beads 2–3 times with extraction buffer.
2. Add 40 μL of 50% TrueBlot anti-rabbit Ig beads (eBioscience) and a few μg of BAK1 or FLS2 antibodies to 10–15 mL crude extract using a cut tip and incubate on a rotator at 4 °C for 2–4 h (*see* **Note 15**).
3. Centrifuge 500× *g* for 1 min at 4 °C.
4. Remove the supernatant using a pipette or a vacuum pump.
5. Add a small amount of extraction buffer and transfer the agarose beads to a new 1.5 mL low-protein binding tube using a 1 mL cut pipette tip.
6. Wash the 50 mL tube with 1 mL of fresh extraction buffer to collect any remaining agarose beads attached to the tube surface and transfer to the 1.5 mL tube.
7. Centrifuge for 30 s at 500× *g* at 4 °C, remove the supernatant, and re-suspend the beads in 1 mL extraction buffer.
8. Wash 2–4 more times with 1 mL extraction buffer.
9. Remove washing buffer and eluted proteins with 50 μL 2xLDS buffer and heated at 70 °C for 15 min (*see* **Note 16**).

10. Perform standard SDS-PAGE and immunoblotting, using 5% skim milk in TBST for blocking.
11. Use TrueBlot Anti-IgG HRP Rabbit (Mouse) as the secondary antibody to prevent interference from heavy and light chains of immunoglobulins.

    Important: If the levels of immunoprecipitated protein (bait) vary, quantify band intensity using ImageJ software (https://imagej.net/ij/) and rerun SDS-PAGE for immunoblotting to ensure equal protein loading.

3.8 SDS-PAGE for LC-MS/MS Analysis

1. Perform SDS-PAGE using commercially available pre-cast gels to ensure consistent band patterns and to remove contaminants from buffers used for IP and elution that may interfere with LC-MS/MS analysis. We use NuPAGE® Bis-Tris Gel (1.5 mM, 10-well), NuPAGE® MOPS SDS Running Buffer, and NuPAGE® Antioxidant.
2. Run the gel at 200 V for 50 min. If a large sample volume is applied, it is recommended to run at a lower voltage (100 V) to ensure better resolution (*see* **Note 17**).
3. Transfer the gel to a new plastic case.
4. Stain the gel with InstantBlue® Coomassie Protein Stain (abcam).
5. Cut bands of interest for LC-MS/MS analysis or cut the entire lane into several pieces, then proceed to trypsin digestion and LC-MS/MS analysis. Consult your collaborator for LC-MS/MS analyses, as the optimal sample preparation method depends on the capabilities of the LC-MS/MS instrument.

4 Notes

1. If you aim to identify phosphorylation sites or assess whether the phosphorylation state influences RLK interactions with associated proteins, include a protein phosphatase inhibitor cocktail and EDTA in the extraction buffer. EDTA chelates Zn^{2+}, leading to the inactivation of Zn^{2+}-dependent metalloproteases. However, some proteins require cations to maintain their proper structure and interactions with binding partners. Additionally, EDTA inhibits protein phosphatase 2C by chelating Mg^{2+} and Mn^{2+}.
2. Protein extraction can be performed by grinding frozen samples with a mortar and pestle, and using sand as an abrasive improves extraction efficiency. For even higher protein yield, homogenization with a POLYTRON homogenizer can be

carried out after manual grinding. Do not use sand if you want to homogenize.

3. You can also adjust the salt concentration based on experimental needs. The normal range is 100–150 mM NaCl, but higher or lower concentrations can be used depending on the objective. Higher salt concentrations reduce non-specific interactions by shielding electrostatic charges but may also weaken transient or ionic interactions. In contrast, lower salt concentrations help preserve weak or transient interactions but can increase background noise due to enhanced non-specific binding and protein aggregation.
4. Higher detergent concentrations may enhance the solubilization of plasma membrane proteins, but they also increase stringency, which can weaken the detection of protein-protein interactions by disrupting weak or transient associations.
5. The expression of a sufficient amount of active RLK fused to a tag with proper localization is crucial for successful IP experiments, whether for identifying phosphorylation sites or novel protein interactors. We recommend generating stable transgenic plants expressing epitope-tagged RLKs using genomic DNA constructs driven by their native promoters. Mislocalized or non-functional RLKs are unlikely to yield biologically relevant interactors. Based on our experience, obtaining several hundred nanograms of RLK after IP is ideal for both isolating interactors and identifying phosphorylation sites.
6. Extracting proteins strongly attached to the plasma membrane can also result in the Co-IP of plasma membrane fragments. Higher detergent concentrations can help disrupt the plasma membrane, but it is crucial to include an appropriate negative control. For example, we currently use the 35S:GFP-LTI6b transgenic line as a negative control for the Co-IP of GFP-fusion proteins localized at the plasma membrane [9]. This is an ideal control because LTI6b is a well-characterized plasma membrane marker that does not interact with PRRs [31].
7. If the goal is to identify specific protein interactors of activated RLKs or their phosphorylation sites, it is crucial to activate as many RLKs as possible in plant cells. If only 10% of RLKs are activated, the remaining 90% of non-activated RLKs will contribute to background noise, significantly reducing the chances of detecting specific interactors or phosphorylation sites. Therefore, it is essential to carefully optimize the concentration of PAMPs and ensure precise vacuum infiltration to achieve robust RLK activation.
8. This protocol can also be used for Co-IP of tagged proteins transiently expressed in *N. benthamiana* via *Agrobacterium*

tumefaciens infiltration. However, overexpressing proteins in confined compartments, such as the plasma membrane or organelles, may result in false interactions. To distinguish specific interactors from artifacts, it is crucial to include appropriate negative controls.

9. The extraction buffer for *N. benthamiana* is the same as for *Arabidopsis*, except that it contains a higher concentration of Tris–HCl (150 mM) and DTT (10 mM). Additionally, PVPP is required to remove phenolic compounds from *N. benthamiana* leaf extracts, helping to prevent protein degradation and reduce non-specific binding.
10. To perform Co-IP on multiple samples simultaneously, grind the samples in advance and store them at −80 °C. This approach allows for the simultaneous addition of extraction buffer to all samples, ensuring consistency in protein extraction and minimizing variability between replicates.
11. If the aim is to identify novel interactors, it is crucial to isolate protein complexes containing RLKs while preserving associated proteins. Excessive washing can help reduce non-specific interactions of sticky proteins, but it may also disrupt the complex of interest, leading to the loss of true interactors. Additionally, over-washing might dissociate ligands from RLKs, resulting in the loss of ligand-dependent interactions.
12. The magnetic bead system enables the rapid isolation of RLK complexes under mild conditions, minimizing the risk of complex dissociation. Magnetic beads from Miltenyi Biotec are extremely small and can be efficiently collected using a μMACS Separator, which applies a magnetic field to retain beads within a column, eliminating the need for centrifugation. This system is particularly well-suited for identifying RLK interactors, though it is important to note that the column's binding capacity is limited. The maximum volume of magnetic bead solution for a μ Column is approximately 250 μL. Various magnetic bead systems are commercially available, and selecting the most appropriate one for the intended purpose is recommended.
13. To identify RLK phosphorylation sites, it is essential to purify a large amount of RLK under conditions that inhibit protein phosphatase activity. Therefore, an IP system that maximizes RLK yield is required. To prevent dephosphorylation, wash the beads multiple times with a buffer containing protein phosphatase inhibitors, ensuring the isolation of pure RLK protein without phosphatase activity. GFP-Trap, an agarose-conjugated antibody, binds strongly to GFP, enabling the purification of high amounts of GFP-fusion proteins. This high-

affinity interaction makes GFP-Trap an effective system for identifying RLK phosphorylation sites.

14. Some membrane proteins are highly unstable at high temperatures, even when mixed with SDS or LDS buffer, and may degrade or precipitate upon boiling. To prevent protein loss, instead of boiling samples in SDS buffer, heat them at a lower temperature (75 °C) for a longer duration (≥15 min). This approach maintains protein stability while ensuring sufficient denaturation for SDS-PAGE.
15. Rabbit antibodies were generated using C-terminal peptide sequences from *Arabidopsis* BAK1 (DSTSQIENEYPSGPR) and FLS2 (KANSFREDRNEDREV). To enhance specificity, the antibodies were antigen-purified using the corresponding peptides.
16. The PRR-BAK1 interaction is very strong, making it difficult to detect without proper denaturation. To ensure detection, boil the samples in SDS buffer for a few minutes or heat them at 75 °C for 15 min.
17. Highly helical proteins migrate faster than expected in SDS-PAGE, running faster than predicted based on molecular size. This effect is more pronounced in higher-percentage gels [32].

Acknowledgments

We thank Yoko Nagai, Noriko Maki, Naomi Watanabe, and Mizuki Yamamoto for providing technical support, and Nobuaki Ishihama for discussions and critical reading of the manuscript. This work was financially supported by Grant-in-Aid for MEXT KAKENHI Grant Numbers 20H02994, 21K19128, and 24K01764 (to Y.K.), JSPS (JP22H00364, JP20H05909), JST (JPMJGX23B2/GteX, JPMJAP2306/ASPIRE), and RIKEN TRIP initiative (to K.S.).

References

1. Shiu SH, Bleecker AB (2001) Receptor-like kinases from Arabidopsis form a monophyletic gene family related to animal receptor kinases. Proc Natl Acad Sci U S A 98:10763–10768
2. Jamieson PA, Shan L, He P (2018) Plant cell surface molecular cypher: receptor-like proteins and their roles in immunity and development. Plant Sci 274:242–251
3. Ngou BPM, Wyler M, Schmid MW, Kadota Y, Shirasu K (2024) Evolutionary trajectory of pattern recognition receptors in plants. Nat Commun 15:308
4. Escocard de Azevedo Manhaes AM, Ortiz-Morea FA, He P, Shan L (2021) Plant plasma membrane-resident receptors: surveillance for infections and coordination for growth and development. J Integr Plant Biol 63:79–101
5. Hailemariam S, Liao CJ, Mengiste T (2024) Receptor-like cytoplasmic kinases: orchestrating plant cellular communication. Trends Plant Sci 29:1113–1130
6. Schwessinger B, Roux M, Kadota Y, Ntoukakis V, Sklenar J, Jones A, Zipfel C (2011) Phosphorylation-dependent differential regulation of plant growth, cell death, and

innate immunity by the regulatory receptor-like kinase BAK1. PLoS Genet 7:e1002046
7. DeFalco TA, Zipfel C (2021) Molecular mechanisms of early plant pattern-triggered immune signaling. Mol Cell 81:4346
8. Roux M, Schwessinger B, Albrecht C, Chinchilla D, Jones A, Holton N, Malinovsky FG et al (2011) The Arabidopsis leucine-rich repeat receptor-like kinases BAK1/SERK3 and BKK1/SERK4 are required for innate immunity to hemibiotrophic and biotrophic pathogens. Plant Cell 23:2440–2455
9. Kadota Y, Sklenar J, Derbyshire P, Stransfeld L, Asai S, Ntoukakis V, Jones JD et al (2014) Direct regulation of the NADPH oxidase RBOHD by the PRR-associated kinase BIK1 during plant immunity. Mol Cell 54:43–55
10. Liebrand TW, van den Berg GC, Zhang Z, Smit P, Cordewener JH, America AH, Sklenar J et al (2013) Receptor-like kinase SOBIR1/EVR interacts with receptor-like proteins in plant immunity against fungal infection. Proc Natl Acad Sci U S A 110:10010–10015
11. Fabregas N, Li N, Boeren S, Nash TE, Goshe MB, Clouse SD, de Vries S et al (2013) The brassinosteroid insensitive1-like3 signalosome complex regulates Arabidopsis root development. Plant Cell 25:3377–3388
12. Karlova R, Boeren S, Russinova E, Aker J, Vervoort J, de Vries S (2006) The Arabidopsis SOMATIC EMBRYOGENESIS RECEPTOR-LIKE KINASE1 protein complex includes BRASSINOSTEROID-INSENSITIVE1. Plant Cell 18:626–638
13. Goto Y, Kadota Y, Mbengue M, Lewis JD, Matsui H, Maki N, Ngou BPM et al (2024) The leucine-rich repeat receptor kinase QSK1 regulates PRR-RBOHD complexes targeted by the bacterial effector HopF2Pto. Plant Cell 36: 4932–4951
14. Macho AP, Schwessinger B, Ntoukakis V, Brutus A, Segonzac C, Roy S, Kadota Y et al (2014) A bacterial tyrosine phosphatase inhibits plant pattern recognition receptor activation. Science 343:1509–1512
15. Bender KW, Couto D, Kadota Y, Macho AP, Sklenar J, Derbyshire P, Bjornson M et al (2021) Activation loop phosphorylation of a non-RD receptor kinase initiates plant innate immune signaling. Proc Natl Acad Sci U S A 118
16. Zipfel C (2014) Plant pattern-recognition receptors. Trends Immunol 35:345–351
17. Zipfel C, Kunze G, Chinchilla D, Caniard A, Jones JD, Boller T, Felix G (2006) Perception of the bacterial PAMP EF-Tu by the receptor EFR restricts Agrobacterium-mediated transformation. Cell 125:749–760
18. Gomez-Gomez L, Boller T (2000) FLS2: an LRR receptor-like kinase involved in the perception of the bacterial elicitor flagellin in Arabidopsis. Mol Cell 5:1003–1011
19. Chinchilla D, Zipfel C, Robatzek S, Kemmerling B, Nurnberger T, Jones JD, Felix G et al (2007) A flagellin-induced complex of the receptor FLS2 and BAK1 initiates plant defence. Nature 448:497–500
20. Heese A, Hann DR, Gimenez-Ibanez S, Jones AM, He K, Li J, Schroeder JI et al (2007) The receptor-like kinase SERK3/BAK1 is a central regulator of innate immunity in plants. Proc Natl Acad Sci U S A 104:12217–12222
21. Schulze B, Mentzel T, Jehle AK, Mueller K, Beeler S, Boller T, Felix G et al (2010) Rapid heteromerization and phosphorylation of ligand-activated plant transmembrane receptors and their associated kinase BAK1. J Biol Chem 285:9444–9451
22. Sun Y, Li L, Macho AP, Han Z, Hu Z, Zipfel C, Zhou JM et al (2013) Structural basis for flg22-induced activation of the Arabidopsis FLS2-BAK1 immune complex. Science 342: 624–628
23. Macho AP, Zipfel C (2014) Plant PRRs and the activation of innate immune signaling. Mol Cell 54:263–272
24. Yeh YH, Panzeri D, Kadota Y, Huang YC, Huang PY, Tao CN, Roux M et al (2016) The Arabidopsis malectin-like/LRR-RLK IOS1 is critical for BAK1-dependent and BAK1-independent pattern-triggered immunity. Plant Cell 28:1701–1721
25. Bucherl CA, Jarsch IK, Schudoma C, Segonzac C, Mbengue M, Robatzek S, MacLean D et al (2017) Plant immune and growth receptors share common signalling components but localise to distinct plasma membrane nanodomains. Elife 6
26. Keinath NF, Kierszniowska S, Lorek J, Bourdais G, Kessler SA, Shimosato-Asano H, Grossniklaus U et al (2010) PAMP (Pathogen-associated Molecular Pattern)-induced changes in plasma membrane compartmentalization reveal novel components of plant immunity. J Biol Chem 285:39140–39149
27. Kadota Y, Macho AP, Zipfel C (2016) Immunoprecipitation of plasma membrane receptor-like kinases for identification of phosphorylation sites and associated proteins. Methods Mol Biol 1363:133–144
28. Mbengue M, Bourdais G, Gervasi F, Beck M, Zhou J, Spallek T, Bartels S et al (2016)

Clathrin-dependent endocytosis is required for immunity mediated by pattern recognition receptor kinases. Proc Natl Acad Sci U S A 113:11034–11039

29. Robatzek S, Chinchilla D, Boller T (2006) Ligand-induced endocytosis of the pattern recognition receptor FLS2 in Arabidopsis. Genes Dev 20:537–542
30. Bourdais G, McLachlan DH, Rickett LM, Zhou J, Siwoszek A, Haweker H, Hartley M et al (2019) The use of quantitative imaging to investigate regulators of membrane trafficking in Arabidopsis stomatal closure. Traffic 20: 168–180
31. Cutler SR, Ehrhardt DW, Griffitts JS, Somerville CR (2000) Random GFP::cDNA fusions enable visualization of subcellular structures in cells of Arabidopsis at a high frequency. Proc Natl Acad Sci U S A 97:3718–3723
32. Rath A, Cunningham F, Deber CM (2013) Acrylamide concentration determines the direction and magnitude of helical membrane protein gel shifts. Proc Natl Acad Sci U S A 110:15668–15673
33. Zhang J, Li W, Xiang TT, Liu ZX, Laluk K, Ding XJ, Zou Y et al (2010) Receptor-like cytoplasmic kinases integrate signaling from multiple plant immune receptors and are targeted by a Pseudomonas syringae effector. Cell Host Microbe 7:290–301

Chapter 3

Plant-Based Ligand-Depletion Binding Assay for Assessing Pattern Recognition Receptor-Ligand Interactions

Lin-Jie Shu, Fan-Yu Yu, and Stefanie Ranf

Abstract

The assessment of cell-surface receptor-ligand interactions is of fundamental importance for understanding plant signal transduction, particularly in immune responses mediated by pattern recognition receptors (PRRs). Iconic biophysical methods such as microscale thermophoresis, isothermal titration calorimetry, and surface plasmon resonance provide precise quantification of binding properties but require specialized equipment, costly reagents, and large quantities of purified proteins or labeled ligands. To offer a more accessible and plant-based approach, this chapter introduces a ligand-depletion binding assay using *Nicotiana benthamiana* to express PRR extracellular domains (ECDs) and *Arabidopsis thaliana* seedlings as sensitive biosensors for detecting ligands. Using the ECD of the PRR LORE and its ligand, 3-hydroxydecanoic acid (3-OH-C10:0), as an example, we demonstrate a workflow in which recombinant PRR ECDs are transiently expressed in the apoplast of *N. benthamiana* via *Agrobacterium*-mediated transformation. The ligand-binding assay utilizes ultrafiltration by concentrator columns to separate ECD-ligand complexes and unbound ligands by molecular weight. *A. thaliana* Ca^{2+} signaling responses serve as a robust readout for ligand detection. This cost-effective and scalable pipeline provides an alternative to conventional methods, enabling medium-throughput screening of receptor variants and ligand candidates. The approach can be broadly adapted to study diverse PRR-ligand interactions, facilitating research in plant immunity.

Key words Ligand-depletion binding assay, Pattern recognition receptor, PAMP/MAMP, LORE, 3-Hydroxy fatty acid, *Agrobacterium*-mediated transient protein expression, Apoplastic washing fluid, Ca^{2+} signaling

1 Introduction

In plants, cell-surface receptor-ligand interactions play a crucial role in regulating diverse physiological processes, such as growth, development, immunity, and environmental responses. Through binding to ligands, cell-surface receptors are activated or inhibited and induce or suppress downstream signaling, resulting in further cellular responses [1]. The specificity of the receptors for their ligands

Tsutomu Kawasaki and Hirofumi Yoshioka (eds.), *Plant Immunity: Methods and Protocols*, Methods in Molecular Biology, vol. 3012,
https://doi.org/10.1007/978-1-0716-5138-4_3,

is critical for accurately detecting target signals and eliciting the appropriate response pathways.

Plant pattern recognition receptors (PRRs) are indispensable cell surface sensors that monitor signals released from plant-colonizing (micro)organisms, including plant pathogens, and induce pattern-triggered immunity (PTI) that effectively controls their colonization and thus fends off most pathogens [2]. Plants detect colonizers by sensing the presence of conserved structural motifs, classically named pathogen-/microbe-associated molecular patterns (PAMPs/MAMPs), as danger signals [3]. For instance, *A. thaliana* senses a peptide motif, flg22, from bacterial flagellin by a leucine-rich-repeat receptor-like kinase-type PRR, FLS2 [4, 5], and chitin fragments released from fungal cell walls are sensed by specific lysin motif-type receptor complexes [6, 7]. The extracellular domain (ECD) structures of PRRs determine their ligand specificity. LORE, comprising a G-type lectin, an epidermal growth factor, and a plasminogen-apple-nematode domain in its ECD, is a PRR sensing medium-chain 3-hydroxy fatty acids mainly released from bacteria [8–10]. The ligand-induced activation of PRRs results in PTI responses, including cytoplasmic Ca^{2+} signaling, reactive oxygen species (ROS) production, accumulation of phytoalexins, and the expression of defense-related genes [11].

Assessing receptor-ligand interactions is crucial for elucidating the relationships between ligand-producing organisms and their corresponding receptor-expressing hosts. Consequently, a variety of techniques are employed to validate these biomolecular interactions [12]. Using advanced and well-established tools, interactions between various PRRs and ligands have been systematically identified, with techniques such as microscale thermophoresis, isothermal titration calorimetry, and surface plasmon resonance as prominent examples [13–15]. While these methods enable the quantification of binding properties, many require substantial amounts of purified receptor proteins or ligand labeling, as well as specialized facilities, costly equipment, or expensive consumables.

In this chapter, taking the LORE ECD and its ligand 3-hydroxy decanoic acid (3-OH-C10:0) as examples, we will demonstrate a workflow established on plant systems, which is more easily accessible and applicable for plant scientists studying biomolecular interactions, named plant-based ligand-depletion binding assay [16]. Most plant receptor recombinant proteins studied to date were expressed in classical heterologous expression systems, including *Escherichia coli* and insect cells [17, 18]. However, these expression approaches do not always yield recombinant proteins with configurations and properties closely resembling their native forms in the original plant species. Utilizing *Nicotiana benthamiana* as an expression host is an effective and reliable way to produce plant proteins, which is widely used in plant research for *in planta* protein studies and has more recently been applied for recombinant protein expression for structural studies, but also for drug and antibody production [19–21]. Most plant PRRs from other plant species studied to date are functional when expressed in *N. benthamiana*, including LORE [8, 9, 22]. We will demonstrate the protocol for *Agrobacterium*-mediated transient expression of soluble PRR ECD protein in the apoplast of *N. benthamiana* leaves and harvesting by extraction of apoplastic washing fluids (AWFs). Next, we will demonstrate the two-step ligand-depletion binding assay, which is based on a simple technique to separate unbound ligands from ECD-ligand complexes by molecular weight using ultrafiltration with protein concentrator spin columns. Finally, to detect the isolated ligands, we will use *A. thaliana* seedlings as biosensors to measure the bioactive ligands in the separated fractions. The 3-OH-C10:0-induced cytoplasmic Ca^{2+} burst serves as a robust and sensitive detector of the presence of the target ligand [9, 23]. When 3-OH-C10:0 is bound by the ECD proteins, the level of 3-OH-C10:0 in the unbound fraction is too low to induce a Ca^{2+} burst (step 1: "depletion assay"), while the bound 3-OH-C10:0, upon release from the ECD proteins by heat denaturing, induces a Ca^{2+} burst (step 2: "binding assay") (Fig. 1). This workflow can easily be adapted for different types of ligands. The ligand-depletion binding assay developed here is cost-effective and suitable for medium-throughput screening of multiple receptor variants or ligand candidates [16]. The concept can be broadly applied to study PRR-ECD-ligand interactions in plants.

2 Materials

2.1 Agrobacterium-*Mediated Transient Protein Expression in the Apoplast of* Nicotiana benthamiana

1. *Agrobacterium tumefaciens* GV3101 (rifampicin resistance) with the Ti helper plasmid pMP90 (gentamycin resistance):
   - Carrying a modified pCB302 binary vector adapted for GoldenGate-cloning containing the target genes (*see* **Note 1**), encoding a LORE-ECD-mCherry fusion protein or a LORE signal peptide-mCherry fusion for targeting mCherry to the apoplast (apo-mCherry, negative control), with kanamycin resistance (*see* **Notes 2** and **3**).
   - Carrying a binary vector with an expression cassette for the viral silencing suppressor p19 (*see* **Note 4**), with kanamycin resistance.
2. *Nicotiana benthamiana*: 6-week-old, grown on a 9:1 mix of potting soil (ED73, Einheitserde, Germany) and fine vermiculite (2–3 mm granulation, Floragard, Germany) in a plant growth chamber under long-day (16-h daylight) conditions at 23 °C/21 °C (day/night).

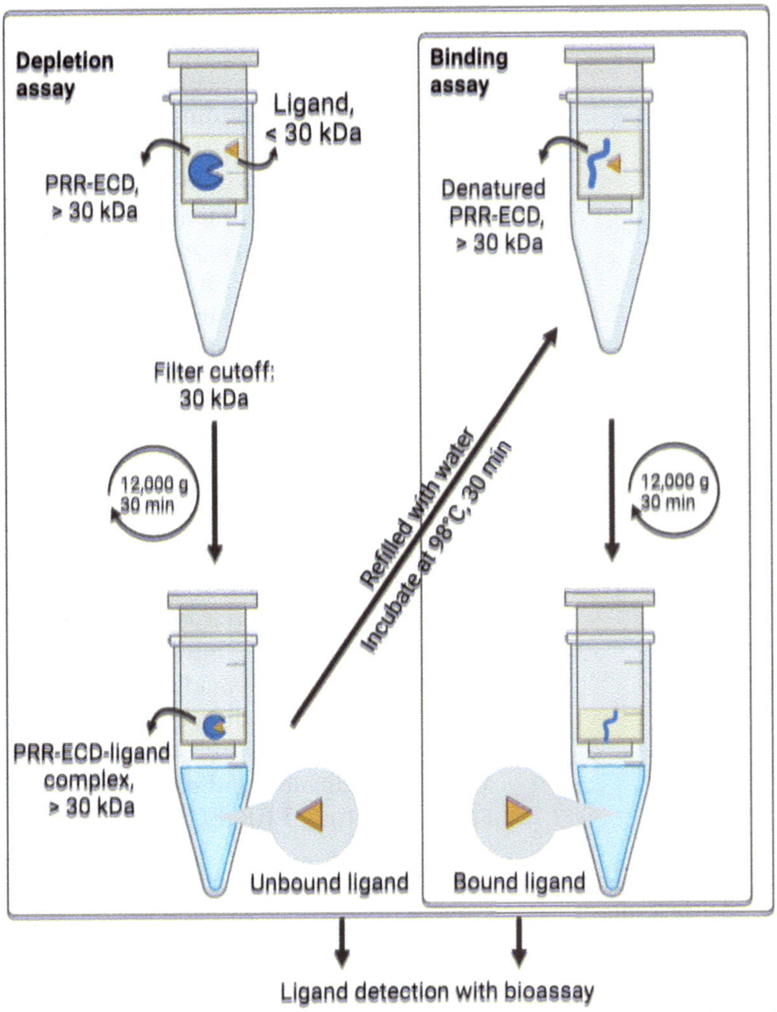

Fig. 1 Schematic representation of the ligand-depletion binding assay. The ligand 3-OH-C10:0 is incubated with concentrated apoplastic washing fluids (AWFs) obtained from *Nicotiana benthamiana* leaves transiently expressing the extracellular domain (ECD) of the pattern-recognition receptor (PRR) of interest. **Depletion assay**: The mixture is subjected to centrifugal ultrafiltration using a 30 kDa molecular weight cut-off membrane. Upon centrifugation, unbound 3-OH-C10:0 (188 Da) passes into the filtrate, while PRR-ECD-ligand complexes are retained. The presence of unbound 3-OH-C10:0 ligand in the filtrate is assessed using a LORE-dependent bioassay; a reduction in signal indicates ligand binding by the PRR-ECD. **Binding assay**: The retentate is refilled to its original volume with water and subjected to heat denaturation at 98 °C for 30 min to release bound ligand. A second ultrafiltration step separates the denatured PRR-ECD protein and the released 3-OH-C10:0, which is recovered in the filtrate and detected by the same bioassay

3. YEB agar (1 g/L yeast extract, 5 g/L beef extract, 5 g/L peptone, 5 g/L sucrose, 0.5 g/L $MgSO_4$-$7H_2O$, 10 g/L agar) with 30 μg/mL gentamycin, 10 μg/mL rifampicin, and 50 μg/mL kanamycin.
4. LB broth (10 g/L peptone, 5 g/L yeast extract, 10 g/L NaCl) with 30 μg/mL gentamycin, 10 μg/mL rifampicin, and 50 μg/mL kanamycin.
5. Acetosyringone (3′,5′-Dimethoxy-4′-hydroxyacetophenone) 100 mM (19.62 mg/mL, in absolute ethanol, stored at – 20 °C) (Merck) (*see* **Note 5**).
6. Freshly prepared infiltration buffer (10 mM $MgSO_4$, 10 mM 2-(N-morpholino)-ethanesulfonic acid (MES) pH 5.5, 150 μM acetosyringone).
7. 15 mL conical centrifuge tube (Falcon).
8. 1 mL needleless plastic syringe (BD®).
9. Refrigerated centrifuge with a fixed angle rotor for 15 mL conical centrifuge tubes (Eppendorf Centrifuge 5910 Ri, Rotor S-4x400 with adaptors for 15 mL conical centrifuge tubes).
10. Spectrophotometer for measuring optical density (OD) at 600 nm (Eppendorf BioPhotometer 6131).

2.2 Harvesting of Nicotiana benthamiana Apoplastic Washing Fluid (AWF)

1. Tris-buffered saline (TBS) buffer (50 mM Tris-base, 150 mM NaCl, pH 7.6).
2. Low (squat)-form 600 mL and high-form 600 mL beakers (Simax).
3. Vacuum pump (PC 3001 VARIO select, Vacuubrand®).
4. Desiccator (Nalgene Classic Design Desiccator with Vacuum Capability from Thermo Fisher Scientific).
5. Paper towels.
6. 1 mL pipette tip.
7. 20 mL needless plastic syringe (BD®).
8. 50 mL conical centrifuge tube (Falcon).
9. Refrigerated centrifuge with a fixed angle rotor for 50 mL conical centrifuge tubes (Eppendorf Centrifuge 5910 Ri, rotor FA-6x50).
10. Liquid nitrogen.
11. Vivaspin 20 centrifugal concentrator with 30 kDa molecular weight cut-off (MWCO) polyethersulfone membrane (Sartorius AG).

12. Protease inhibitor cocktail for plant cell and tissue extracts, dissolved in DMSO and stored at −20 °C (Sigma-Aldrich, P9559).
13. Ultra-pure sterile water.
14. PD-10 desalting columns with Sephadex G-25 resin (Cytiva).

2.3 Quality Assessment and Quantification of the Proteins in the AWF

1. Bradford assay dye reagent concentrate and BSA protein standard (Protein Assay Kit I, Bio-Rad).
2. 96-well clear flat-bottom microplate (Greiner Bio-One International GmbH).
3. 12-channel pipette, 20–200 μL (Eppendorf Research® plus).
4. Microplate spectrophotometer for measuring absorbance at 595 nm for Bradford protein assay (Tecan Infinite 200 Pro plate reader).
5. 5x Laemmli sample buffer (60 mM Tris–HCl pH 6.8, 10% glycerol, 2% SDS, 5% β-mercaptoethanol, 0.01% bromophenol blue).
6. Mini gel electrophoresis system with power supply (Bio-Rad Mini-PROTEAN Tetra Vertical Electrophoresis Cell and PowerPac Universal Power Supply).
7. SurePAGE Bis-Tris 10% (GenScript).
8. Tris-MOPS-SDS Running Buffer (GenScript).
9. PageRuler Prestained Protein Ladder (Thermo).
10. RAPIDstain blotting-compatible Coomassie Brilliant Blue staining (G-Biosciences).
11. 0.2 μM nitrocellulose western blotting membrane (Cytiva).
12. Semi-dry blotting system (Bio-Rad Trans-Blot SD Semi-Dry Transfer Cell).
13. 1x transfer buffer (3.03 g/L Tris base, 14.4 g/L glycine, 20% methanol, and 0.05% SDS).
14. Protein-free blocking buffer (Serva).
15. Primary antibody anti-RFP, clone 5F8 (Proteintech).
16. TBS-T buffer (TBS buffer, pH 7.6, 0.05% Tween 20).
17. Secondary antibody anti-rat-horseradish peroxidase (HRP) conjugate A9542 (Merck).
18. SuperSignal West Dura Extended Duration Substrate (Thermo).
19. Image acquisition system for the capture of chemiluminescent images (Fusion-SL4, Vilber Lourmat).

2.4 Ligand-Depletion Binding Assay

1. 3-hydroxy decanoic acid (3-OH-C10:0; Cayman Chemical, # 24613), 5.31 mM (1 mg/mL) in absolute ethanol (analytical grade), stored at −20 °C (*see* **Note 6**).
2. Protein LoBind tubes, 1.5 mL (Eppendorf).
3. Rotator (DLAB MX-RD-E Economical Rotator).
4. Vivaspin 500 centrifugal concentrator with 30 kDa MWCO polyethersulfone membrane (Sartorius AG).
5. Refrigerated centrifuge with fixed angle rotor for microtubes (Eppendorf Centrifuge 5425 R, rotor FA-30x2).
6. Microtube (Axygen® 1.7 mL MaxyClear Snaplock Microcentrifuge Tube, ref. no. MCT-175-C) or PCR tube (Sapphire 8-tube Strips with individually attached caps 0.2 mL, Greiner Bio-One, ref no. 673283).
7. Dry bath (Techne Dri-Block DB-2D Digital Blockheater) or PCR thermocycler (GeneExplorer Thermal Cycler, model GE-48D).

2.5 Measurement of Cytosolic Calcium Burst Induced by the Unbound and Bound Fractions

1. *Arabidopsis thaliana* Col-0 expressing apoaequorin under control of the cauliflower mosaic virus (CaMV) 35S promoter in the cytosol (Col-0^{AEQ}) [23].
2. 24-well clear flat-bottom cell culture plate, with lid, sterile (Greiner Bio-One International GmbH).
3. Breathable sealing film Axygen BF-400-S (Corning).
4. Glass desiccator (DURAN® Vacuum Desiccator, NOVUS type).
5. Sodium hypochlorite solution 13–14% ($REACTOLAB_{SA}$, ref. no. 99412).
6. 37% Hydrochloric acid, AR-certified for analysis, $d = 1.18$ (Fisher Chemical™).
7. Half-strength Murashige & Skoog liquid medium (2.2 g/L Murashige & Skoog medium including vitamins (Duchefa Biochemie, Netherland, M0222), 2.5 g/L sucrose, 0.195 g/L MES, pH 5.7).
8. Precision tweezers (Dumont 0108-5-PO Tweezers NO.5 Inox 08).
9. Microplate luminometer (Luminoskan Ascent, Thermo Fisher Scientific).
10. Microplate LUMITRAC200, polystyrene, 96 well, F-bottom (chimney well), white, medium binding (Greiner Bio-One International GmbH).
11. 10 mM coelenterazine-h (p.j.k. GmbH) (*see* **Note 7**), dissolved in absolute ethanol and stored in the dark at −80 °C.
12. Discharging solution (2 M $CaCl_2$, 20% ethanol).

13. 12-channel pipette, 5–100 μL (Eppendorf Research® plus).
14. Micropore tape (3M™ Micropore™ Surgical Tape 1530-1, 1 IN x 10 YD (2.5 cm × 9.1 m)).

3 Methods

3.1 Agrobacterium-Mediated Transient Protein Expression in the Apoplast of Nicotiana benthamiana

1. Grow *A. tumefaciens* strains, each carrying one of the protein expression vectors for LORE-ECD-mCherry fusion protein or apo-mCherry or the silencing suppressor p19, on YEB agar with suitable antibiotics and 100 μM acetosyringone at 28 °C for 36–48 h.
2. Sub-culture agrobacteria in 3 mL of LB broth in 15 mL centrifuge tubes with suitable antibiotics and 100 μM acetosyringone at 28 °C, 200 rpm overnight.
3. Collect agrobacteria from the overnight cultures directly in the culture tube by centrifuging at 3000× *g* for 5 min at room temperature. Then, resuspend the bacterial pellet in 3 mL of the freshly prepared infiltration buffer.
4. Wash the bacterial pellet by centrifuging at 3000× *g* for 5 min at room temperature and resuspending with 3 mL of fresh infiltration buffer. Repeat this washing step.
5. Measure the OD_{600} of agrobacteria in a spectrophotometer and adjust to 0.5 with infiltration buffer.
6. Incubate the agrobacteria suspension at 28 °C without shaking for 2–3 h.
7. Before infiltration, mix each *Agrobacterium* strain carrying one of the genes of interest with an equal volume and OD_{600} of agrobacteria carrying the silencing suppressor p19.
8. Using a needleless syringe, infiltrate the agrobacteria mixtures into the apoplastic space of fully expanded leaves of 6-week-old *N. benthamiana* plants until most of the leaf area appears moist and exhibits a darker coloration. Infiltrate the abaxial side of the leaf. Label the infiltrated leaves with a tag describing the sample.
9. After agrobacteria infiltration, leave the *N. benthamiana* plants at room temperature until the leaf interior is dry, as indicated by returning to their normal coloration. Then, move the plants back to the growth chamber.

3.2 Harvesting of N. benthamiana Apoplastic Washing Fluid (AWF)

Recombinant proteins accumulated in the apoplast of *N. benthamiana* leaves can be harvested by extracting AWFs (*see* **Note 8**) at 3 days post-inoculation (*see* **Note 9**). Perform the protocol below for each set of leaves infiltrated with a different agrobacteria mixture separately.

1. Fill a beaker with cold TBS buffer. Approximately 150 mL of TBS buffer is required for a 600 mL beaker (*see* **Note 10**).
2. Detach the agro-infiltrated leaves from the petiole, with the petiole left as short as possible to prevent damaging the leaves.
3. Submerge the detached leaves in the beaker containing the TBS buffer. Gently press all leaves into the buffer. This can be achieved by inserting a high-form 600 mL beaker inside the low-form 600 mL beaker containing the TBS buffer (Fig. 2a) (*see* **Note 11**). Gently shake to remove any bubbles trapped between the leaves.

Fig. 2 Procedure for harvesting apoplastic washing fluid (AWF) from *Nicotiana benthamiana* leaves. (**a**) Agro-infiltrated leaves are submerged in approximately 150 mL of TBS buffer (pH 7.6) within a low-form 600 mL beaker placed inside a vacuum desiccator. A second, high-form 600 mL beaker presses the leaves down to ensure full submersion. (**b**) Representative *N. benthamiana* leaves before (left) and after (right) infiltration with TBS buffer. Note the darker coloration post-infiltration. (**c**) An illustration of rolling up leaves around a 1 mL pipette tip. (**d**) Following infiltration, leaves are rolled around a 1 mL pipette tip, placed into a 20 mL syringe barrel, and inserted into a 50 mL centrifuge tube. (**e**) AWF is collected after centrifugation of the setup in (b) at 800× *g* for 15 min at 4 °C. The collected AWF at the bottom of the 50 mL tube is indicated by the black arrow

4. Put the beaker-leaf stack into a vacuum desiccator and draw the vacuum to 100 mbar for 1.5 min.
5. Within 10 s, release the vacuum in the desiccator. The infiltrated leaves will appear moist and darker in color (Fig. 2b).
6. Gently remove the leaves from the beaker, place them on a clean paper towel, and carefully remove the surface moisture on both sides of the leaves.
7. Roll one or two leaves around a 1 mL pipette tip to form a column (Fig. 2c) and put it into a 20 mL syringe barrel without its plunger. Place the syringe barrel with the leaf roll into a 50-mL centrifuge tube (Fig. 2d).
8. Centrifuge the whole assembly for 15 min at 800× *g*, 4 °C, in a fixed-angle rotor. The AWF collects at the bottom of the 50-mL centrifuge tube (Fig. 2e). Keep the collected AWF on ice.
9. Centrifuge the AWF for 20 min at 12,000× *g*, 4 °C. Transfer the cleared supernatant to a fresh tube and keep it on ice. AWFs from leaves expressing the same protein of interest can be pooled.
10. Add protease inhibitor cocktail to the AWFs to a final concentration of 0.1% (*see* **Note 12**).
11. If not processed immediately, the AWFs can be flash-frozen in liquid nitrogen and stored at −20 °C.
12. Desalt the AWFs using PD-10 desalting columns with sterile water, following the manufacturer's instructions (*see* **Note 13**).

3.3 Quality Assessment and Quantification of the Proteins in the AWF

Check the expression of the proteins of interest by SDS-PAGE and immunoblot before performing the ligand-depletion binding assay (Fig. 3).

1. Measure the total protein concentration of the AWF by Bradford protein concentration assay in a 96-well microtiter plate at 595 nm by a microplate spectrophotometer, following the manufacturer's instructions (*see* **Note 14**). Use BSA protein in the range 0.05–1.5 mg/mL as a standard.
2. Adjust the concentrations of the AWF containing LORE-ECD-mCherry and apo-mCherry proteins to 1.5 mg/mL using sterile water in 1.5 mL protein LoBind tubes. If necessary, the AWFs can be concentrated to the desired total protein concentrations using Vivaspin 20 concentrators with 30 kDa MWCO following the manufacturer's instructions (*see* **Note 15**).
3. Mix 7.5 μg of each concentrated AWF protein with 4 μL of 5× Laemmli sample buffer to a total volume of 20 μL and incubate the protein mixture at 95 °C for 10 min.

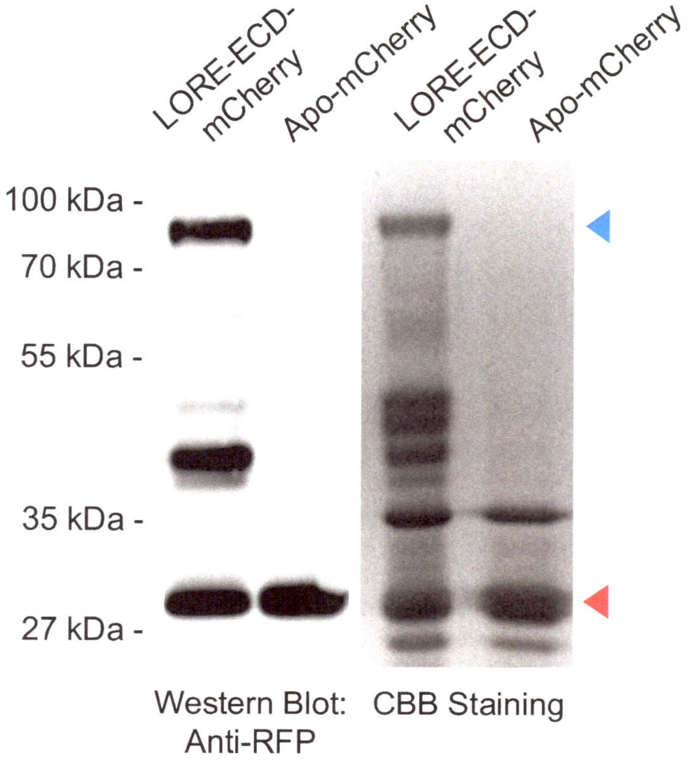

Fig. 3 Analysis of apoplastic washing fluids (AWFs) of *Nicotiana benthamiana* leaves transiently expressing LORE-extracellular domain (ECD)-mCherry fusion protein or apoplastic mCherry (apo-mCherry). Anti-RFP immunoblot (left panel) of desalted and concentrated AWFs from *N. benthamiana* leaves expressing the LORE-ECD-mCherry fusion protein and apo-mCherry. AWFs corresponding to 7.5 μg total proteins were loaded. The theoretical molecular weights of LORE-ECD-mCherry fusion protein and apo-mCherry are 73.9 kDa (blue arrowhead) and 26.7 kDa (red arrowhead), respectively. Equal loading of the samples was confirmed by Coomassie Brilliant Blue (CBB) staining of the SDS-polyacrylamide gel before blotting (right panel). One representative immunoblot of three experiments is shown

4. Load the protein mixture and the PageRuler Prestained Protein marker in a SurePAGE Bis-Tris 10% gel.
5. Run the electrophoresis at 120 V in Tris-MOPS-SDS running buffer until the bromophenol blue in every sample reaches the bottom of the gel.
6. Wash the resolving gel with water for 10 min, then stain for 1 h with RAPIDstain Coomassie brilliant blue staining. Destain the gel with water for 1 h and image (Fig. 3) (*see* **Note 16**).
7. Transfer the proteins from the resolving gel onto a 0.2 μm nitrocellulose western blotting membrane by a semi-dry blotting system with 1x transfer buffer at 1 mA/cm^2 for 70 min.

8. Incubate the nitrocellulose membrane in protein-free blocking buffer at room temperature for 1 h.
9. Incubate the membrane with the primary antibody, anti-RFP (1:2000 dilution in protein-free blocking buffer), for 1 hour at room temperature or 4 °C overnight.
10. Wash the membrane three times for 10 min with TBS-T buffer.
11. Incubate the membrane with the secondary antibody, anti-rat-HRP (1:20,000 dilution, diluted in TBS-T buffer), for 1 hour at room temperature.
12. Wash the membrane three times for 10 min with TBS-T buffer.
13. After removing the TBS-T buffer, incubate the membrane with SuperSignal™ West Dura Extended Duration Substrate for 5 min at room temperature in the dark, following the manufacturer's instructions.
14. Capture the chemiluminescent signal on the membrane by a chemiluminescence image acquisition system, following the manufacturer's instructions (Fig. 3).

3.4 Ligand-Depletion Binding Assay

The ligand-depletion binding assay includes both a "depletion assay step" and a "binding assay step" (Fig. 1). The depletion assay examines whether the PRR ECDs remove the target ligands from the mixture, while the binding assay determines the amount of ligand bound by the PRR ECDs. To ensure sufficient depletion of ligands by the PRR ECD proteins, the molar concentration of the ECD proteins must significantly exceed that of ligands (*see* **Note 17**). The amount of the LORE-ECD-mCherry protein in this protocol is sufficient to deplete the 3-OH-C10:0 ligand at the concentration used below the detection limit.

1. Mix 144 μL of each AWF protein sample with 16 μL of 5 μM 3-OH-C10:0 ligand or another 9:1 (v:v) ratio. Water (without protein) mixed with the same amount of 3-OH-C10:0 is used as a negative control (*see* **Note 18**).
2. Place the mixture on an overhead rotator at 4 °C for at least 1 h to allow ligand binding to reach equilibrium (*see* **Note 19**).
3. For each protein-ligand combination and control, rinse two 30 kDa MWCO Vivaspin 500 concentrators according to the manufacturer's instructions.
4. Transfer the complete AWF-ligand mixture into a rinsed 30 kDa MWCO Vivaspin 500 concentrator. The total volume of the mixture in the concentrator must not exceed 200 μL.
5. Centrifuge the concentrator for 30 min at 12,000× *g*, 4 °C. The 30 kDa MWCO concentrators allow 3-OH-C10:0 to flow through, but proteins/protein-ligand complexes larger than

30 kDa are retained. Hence, the unbound ligand is found in the filtrate, and ECD-ligand complexes are found in the retentate.

6. Refill the filtrate containing unbound ligands (named "unbound fraction") with sterile water up to 160 μL or the original volume of the mixture and store at 4 °C or any other suitable condition to maintain the ligands stable before measurement (*see* **Note 20**).
7. Transfer the retentate into a microtube or PCR tube and refill with sterile water up to 160 μL or the original volume of the mixture.
8. Incubate the retentate in a microtube or a PCR tube in a dry bath or PCR machine at 98 °C for 30 min to denature the proteins and release the ligands.
9. Cool down the retentate on ice and transfer the mixture to a new rinsed 30 kDa MWCO Vivaspin 500 concentrator.
10. Centrifuge the concentrator for 30 min at 12,000× *g*, 4 °C. The released ligands are found in the filtrate, and denatured ECD proteins are found in the retentate.
11. Refill the second filtrate containing the released ligands (named "bound fraction") with sterile water up to 160 μL or the original volume of the mixture and store at 4 °C or any other suitable condition to maintain the ligands stable before measurement (*see* **Note 20**).

3.5 Measurement of Calcium Burst Induced by Ligands in Unbound and Bound Fractions

To detect 3-OH-C10:0 in the unbound and bound fractions from the ligand-depletion binding assays, Col-0^{AEQ} seedlings are used as biosensors to measure the 3-OH-C10:0-induced $[Ca^{2+}]_{cyt}$ burst (*see* **Note 21**). If the unbound fraction does not induce a Ca^{2+} burst, it indicates that most 3-OH-C10:0 molecules are bound by LORE-ECD-mCherry proteins. Vice versa, in the bound fraction, 3-OH-C10:0 molecules released from LORE-ECD-mCherry proteins upon heat denaturation induce a Ca^{2+} burst. Preparation of Col-0^{AEQ} seedlings must be started 10 days before performing ligand-depletion binding assays to avoid prolonged storage of the unbound and bound fractions, which might negatively affect the ligand integrity.

3.5.1 Prepare *A. thaliana* Col-0^{AEQ} Seedlings

1. Remove and store the lid from a sterile 24-well plate under sterile conditions. Distribute sieved *A. thaliana* Col-0^{AEQ} seeds into the wells of the 24-well plates, 15–20 seeds per well, and seal the plate with a gas-permeable breathable seal.
2. Place the sealed plate in a glass desiccator under a chemical fume hood with ventilation. To surface-sterilize the seeds, chlorine gas is generated by mixing 20 mL sodium hypochlorite solution and 10 mL hydrochloric acid, or another 2:1 (v:v)

ratio, in a glass dish placed at the bottom of the desiccator. Immediately close the dessicator tightly with a lid and incubate the plate with the chlorine gas for 3 h.

3. Remove the lid from the desiccator under the chemical fume hood and allow the chlorine gas to evaporate overnight (*see* **Note 22**).
4. The next day, under sterile conditions, carefully remove the breathable seal and add 2 mL of liquid half-strength MS medium in each well. Close the plate with the stored lid and seal it with micropore tape.
5. Stratify the seeds in the dark at 4 °C for at least 2 days.
6. Place the plate in a plant growth chamber under long day (16-h daylight) conditions at 23 °C/21 °C (day/night) 8–9 days before the Ca^{2+} burst measurement.

3.5.2 Prepare Measuring Plate

1. On the day before the measurement, add 75 μL sterile water to each well of a white 96-well microplate and carefully transfer the *A. thaliana* Col-0^{AEQ} seedlings individually from the 24-well plate into the wells with precision tweezers.
2. Add 25 μL of 40 μM coelenterazine-h in each well to reach a final concentration of 10 μM coelenterazine-h. Avoid strong light as coelenterazine-h is light-sensitive (*see* **Note 23**).
3. Cover the 96-well plate with a suitable lid to reduce water evaporation while facilitating oxygen exchange, e.g., the lid of the 24-well plate. Keep the plate overnight in the dark at room temperature or in the plant growth chamber to allow reconstitution of aequorin (*see* **Note 23**).

3.5.3 Measurement of Cytosolic Ca^{2+} Burst

1. Gently transfer the plate into the luminometer and avoid exposure to strong light during placement.
2. One row of the 96-well plate, up to 12 seedlings, is measured in parallel, and the measurement is repeated for eight rounds until the entire plate has been analyzed (*see* **Note 24**).
   (a) Measure the luminescence corresponding to the seedlings' resting $[Ca^{2+}]_{cyt}$ level in 10-s intervals for 2 min in kinetic mode.
   (b) During this time, prepare a PCR strip tube with 12 tubes corresponding to the 12 seedlings to be measured in this round. In each tube, pipette 25 μL of the unbound or bound fraction, respectively, to be tested in this round.
   (c) Move the plate out of the luminometer.
   (d) Treat the seedlings with 25 μL of the unbound or bound fraction using a 12-channel pipette. Immediately resume luminescence measurements for this row for 30 min with 10-s intervals in kinetic mode.

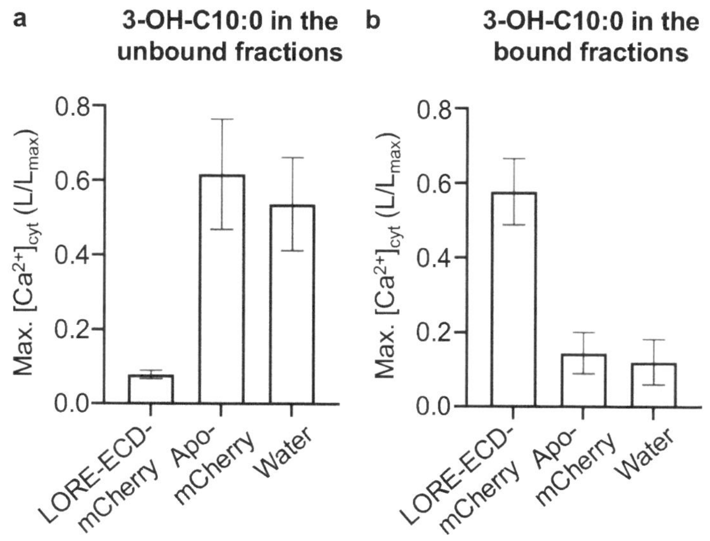

Fig. 4 Example of measuring Ca^{2+} bursts induced by the 3-OH-C10:0-depleted and 3-OH-C10:0-bound fractions. (**a**) 3-OH-C10:0 in the unbound fraction of LORE-ECD-mCherry did not induce a Ca^{2+} burst above the resting cytosolic Ca^{2+} level, whereas the unbound fractions of apoplastic mCherry (apo-mCherry) and water did, indicating that LORE-ECD depleted 3-OH-C10:0 from the mixtures. (**b**) 3-OH-C10:0 in the bound fraction of LORE-ECD-mCherry induced a significant Ca^{2+} burst above the water control, whereas the bound fractions of apo-mCherry did not, indicating that LORE-ECD bound 3-OH-C10:0 from the mixtures. The ligand-depletion binding assay was performed using 500 nM 3-OH-C10:0 and concentrated apoplastic washing fluids (AWFs, 1.5 mg/mL total protein concentration) from *N. benthamiana* containing the LORE-ECD-mCherry fusion proteins or apo-mCherry as a control protein. Water was used as a negative control for the assays. Ca^{2+} bursts in *A. thaliana* Col-0^{AEQ} seedlings, reconstituted with 10 μM coelenterazine-h overnight in water, were measured for 30 min upon application of the unbound and bound fractions from (**a**) the depletion assay and (**b**) the binding assay, respectively. Graphs show means ± s.d. of the maximum cytosolic $[Ca^{2+}]$ levels from three seedlings per fraction

(e) The seedlings are treated individually, one after the other, with 150 μL discharging solution using automatic injection, and the total luminescence counts remaining (L_{max}) are measured for 2.5 min in 10-second intervals in monitor mode (*see* **Note 25**).

3. For each measuring time point, luminescence is normalized to the total luminescence counts remaining (L/L_{max}) (*see* **Note 25**) (Fig. 4).

4 Notes

1. Different binary vector systems for *Agrobacterium*-mediated transient expression in *N. benthamiana* are widely available,

such as pEAQ-HT [24], pICSL4723 [25], and similar vectors. Users should choose one that is compatible with the cloning strategy available in their lab, and the *Agrobacterium* strain and helper plasmid used.

2. To target the proteins of interest to the apoplast, they must either contain their native signal peptide or be fused to a suitable signal peptide of another secreted protein (*see* **Note 3**). The LORE-ECD, the 3-OH-C10:0-binding protein, naturally contains a signal peptide. The cDNA of the LORE-ECD (amino acids 1-424, including its signal peptide) is fused in frame with the downstream coding sequence of a C-terminal-located ten-glycine linker and an mCherry epitope tag (LORE-ECD-mCherry). Similarly, the LORE signal peptide (amino acids 1-21) is fused N-terminally to mCherry to target it to the apoplast to serve as a negative control (apo-mCherry). The coding sequences of LORE-ECD-mCherry and apo-mCherry are each inserted into a binary expression vector between the CaMV 35S promoter and 35S terminator. For detailed information on vectors and cloning, see [16].

3. Signal peptides can significantly influence the transient expression of the gene of interest. Signal peptides derived from the target proteins themselves or other apoplastic proteins, such as *A. thaliana* PR1 (PATHOGENESIS-RELATED GENE 1, AT2G14610), are reasonable choices for optimizing expression. Also, codon usage optimization may be considered to improve the expression of the protein of interest in *N. benthamiana* [21].

4. The tombusvirus p19 suppressor of post-transcriptional gene silencing can significantly boost the expression of genes of interest in *N. benthamiana* [26]. In some cases, the p19-expressing cassette is integrated into a binary vector for expressing the protein of interest, such as pEAQ-HT, to avoid growing two agrobacteria strains each time.

5. Depending on the manufacturer, pure acetosyringone should be stored at −20 °C, 4 °C, or room temperature. However, its stock in organic solvent is best placed at −20 °C for long-term storage.

6. The solubility of 3-OH-C10:0 varies with temperature. Before opening, allow the fatty acid solution to warm to room temperature and sonicate the solution to dissolve any precipitates and mix evenly.

7. Coelenterazine-h is a light-sensitive substrate that can be dissolved in absolute ethanol or methanol. The stock solution should be kept dark at −80 °C and used within a few months. Working aliquots can be stored in the dark at −20 °C for short periods.

8. In the secretory pathway, the proteins of interest undergo post-translational modifications, such as glycosylation, and pass the endoplasmic reticulum quality control system. Harvesting proteins from the apoplast thus yields a higher proportion of mature, properly glycosylated, and correctly folded proteins than extraction from total leaf material. In this protocol, the AWF is suitable for direct use in the LORE-3-OH-FA ligand-depletion binding assay. Protein purification after harvesting AWFs should be considered if the subsequent ligand-binding assays require quantitative analysis, high protein purity, or if the conditions in AWFs are incompatible, for example, because the ligand is degraded by apoplastic enzymes or components in the AWF interfere with the subsequent ligand detection assay. For detailed protocols for protein purification, see [20–22].
9. Theoretically, the longer *N. benthamiana* expresses the target protein, the higher the total yield. However, proteins are degraded by various proteases at different rates within the apoplast already before the harvesting of AWFs [27]. As a result, prolonged expression leads to an increased accumulation of fragmented target proteins at the time of harvest. Therefore, the optimal harvest time must be established for each target protein depending on its stability in the apoplast, the acceptable proportions of fragmentation, and the desired yield.
10. The buffer conditions for harvesting AWF can be adapted based on the properties of the target protein and the requirements of subsequent assays. For example, some studies use a phosphate-buffered saline (PBS)-based buffer instead of TBS buffer, as TBS may interfere with certain in vivo assays. Since the natural pH of the plant apoplast is acidic, acidic buffers may stabilize pH-sensitive proteins.
11. The container for the buffer and leaves can be chosen based on preference, as long as the entire stack fits inside the vacuum desiccator and the leaves do not float on the surface of the buffer.
12. Note that protein degradation occurring in planta before harvesting cannot be prevented, but using ice-cold buffers and protease inhibitors diminishes further degradation of the target protein during and after harvest of AWFs. Check beforehand that the protease inhibitor cocktail does not interfere with the subsequent ligand detection assay. The concentration of the protease inhibitor cocktail can be increased up to 1%. As some low molecular weight inhibitors may be lost by passing through the desalting column and concentrator, it may be necessary to resupply the protease inhibitors in each step.

13. In the plant-based ligand depletion binding assay, *A. thaliana* seedlings are used as biosensors for ligand detection. Since higher concentrations of salts interfere with the Ca^{2+} burst measurement, AWFs must be desalted. Generally, neutral pure water is not favorable for protein stability and solubility. To preserve the target protein for a longer period, storage in an appropriate buffer solution is recommended. For salt-sensitive experiments, such as the cytosolic Ca^{2+} burst assay in this protocol, the buffer should be replaced with pure water, and the protein samples should be used promptly. The buffer in AWFs can be exchanged for other buffers as needed, depending on the specific requirements of subsequent binding assays and ligand measurements

14. Common protein assays, such as the Bradford, BCA, and Lowry, can be considered. These assays have different features, so the appropriate assay should be selected based on the compatibility with the protein sample. Alternatively, protein concentration can be measured spectrophotometrically.

15. The MWCO is selected based on the molecular weight of the proteins of interest. The theoretical molecular weights of LORE-ECD-mCherry fusion protein and apo-mCherry are 73.9 kDa and 26.7 kDa, respectively. The manufacturer recommends an MWCO at least 50% lower than the molecular weight of the target protein. However, it is still possible to concentrate proteins in the same size range as the MWCO, as is the case in this protocol for apo-mCherry. Overall, concentrators should be tested experimentally before large-scale tests are carried out.

16. Relative quantitation, named densitometry, measures the amount of one protein relative to another standard protein using a standard curve based on known concentrations in the same stained SDS-PA gel. For example, BSA in varying amounts can serve as a standard to specifically quantify the proteins of interest instead of measuring the total protein concentration of the AWF, especially when the expression of the proteins of interest to be compared is highly variable. For further information on the relative quantification of proteins in non-purified samples, see [28, 29].

17. To adapt this workflow to different ECD proteins and ligands, the absolute amount of ligand and the molar ratio of ECD protein:ligand for sufficient depletion of ligands need to be estimated and tested experimentally. As a starting point, we recommend a ligand concentration that elicits a robust but non-maximal response in the subsequent ligand detection assay and at least tenfold molar excess of ECD protein over the ligand.

18. The small concentrators used in the ligand-depletion binding assay can hold up to 500 μL. However, they may not effectively filter mixtures with a high protein concentration (1.5 mg/mL in the assay) when the volume exceeds 200 μL. Maintaining a low mixture volume ensures that free ligands can pass through the membrane. If ligand measurement after the binding assay requires large sample volumes, the protein filtration capacity should be carefully evaluated.
19. The time required to reach binding equilibrium may vary depending on the receptor-ligand pair, temperature, and buffer conditions. Initially, overnight ligand binding may be considered. In this case, binding should be performed at low temperatures to slow down protein degradation, especially when protein samples are impure and contain proteases such as the AWFs in this protocol.
20. The retentate cannot be infinitely concentrated because the concentrator has a dead stop volume, meaning it cannot be reduced beyond this limit. The dead stop volume varies depending on the sample type and concentration, operating temperature, and centrifuge rotor. Thus, retentate volumes may vary among different samples. It is necessary to measure the retentate volume first and then adjust it accordingly with water in the binding assay.
21. Any sensitive bioassay in any plant species can be used for detecting bioactive ligands in the unbound and bound fractions. In case ligand variants of interest do not elicit a response in the bioassay, another way of ligand detection in the unbound and bound fractions needs to be chosen, e.g., conjugation with a fluorescent dye or radiolabelling, since the bioassay protocol described here relies on the ligand's bioactivity for its detection.
22. *A. thaliana* seeds can be sterilized using 75% ethanol and bleach if safety requirements prevent using chlorine gas.
23. Aequorin is a Ca^{2+}-sensitive photoprotein consisting of an apoaequorin protein and the luciferin coelenterazine as a prosthetic group. In the presence of oxygen, the holoaequorin spontaneously reconstitutes. Binding of three Ca^{2+} ions to aequorin induces a conformational change that enables the oxidation of coelenterazine to excited coelenteramide, accompanied by the release of CO_2. The relaxation of coelenteramide to its ground state results in the emission of blue light. This bioluminescent signal is readily detectable with a luminometer. In addition to the native coelenterazine, derivatives with higher sensitivity and light output are available, such as coelenterazine-h [30, 31].

24. The Luminoskan Ascent program parameters for the measurement of cytosolic Ca^{2+} burst used in this protocol:
    (i) Background (2 min): measure type: kinetic, integration-time: 240 ms, interval: 10 s, measurement count: 12 times.
    (ii) Elicitation (30 min): measure type: kinetic, integration-time: 240 ms, interval: 10 s, measurement count: 180 times.
    (iii) Discharge (2.5 min): measure type: monitor, integration-time: 10,000 ms, measurement count: 15 times. In this step, each seedling needs to be individually treated and immediately measured to facilitate recording all luminescence readings before moving on to the next seedling in the row.

    For all steps, the PMT voltage is set to 1200 volts.
25. Normalizing the luminescence counts at each time point (L) to the total luminescence counts remaining (L_{max}) compensates for differences in seedling size, apoaequorin expression, and aequorin reconstitution. However, if the luminometer's technical setup does not allow for automatic injection of the discharging solution, this step can also be omitted. In this case, seedlings of comparable size and developmental stage should be selected to allow comparability of results.

Acknowledgments

This work was supported by funds from the German Research Foundation (grants SFB924/TP-B10 and Emmy Noether programme RA2541/1 to SR) and the Swiss National Science Foundation (grant 310030_208139 to SR). Schemes were created with BioRender.com.

References

1. Hohmann U, Lau K, Hothorn M (2017) The structural basis of ligand perception and signal activation by receptor kinases. Annu Rev Plant Biol 68:109–137
2. Ngou BPM, Ding P, Jones JDG (2022) Thirty years of resistance: zig-zag through the plant immune system. Plant Cell 34:1447–1478
3. Shu L-J, Kahlon PS, Ranf S (2023) The power of patterns: new insights into pattern-triggered immunity. New Phytol 240:960–967
4. Felix G, Duran JD, Volko S, Boller T (1999) Plants have a sensitive perception system for the most conserved domain of bacterial flagellin. Plant J 18:265–276
5. Gómez-Gómez L, Boller T (2000) FLS2: an LRR receptor–like kinase involved in the perception of the bacterial elicitor flagellin in Arabidopsis. Mol Cell 5:1003–1011
6. Cao Y, Liang Y, Tanaka K, Nguyen CT, Jedrzejczak RP, Joachimiak A, Stacey G (2014) The kinase LYK5 is a major chitin receptor in Arabidopsis and forms a chitin-induced complex with related kinase CERK1. eLife 3:e03766
7. Miya A, Albert P, Shinya T, Desaki Y, Ichimura K, Shirasu K, Narusaka Y, Kawakami N, Kaku H, Shibuya N (2007) CERK1, a LysM receptor kinase, is essential

for chitin elicitor signaling in Arabidopsis. Proc Natl Acad Sci U S A 104:19613–19618

8. Ranf S, Gisch N, Schäffer M, Illig T, Westphal L, Knirel YA, Sánchez-Carballo PM, Zähringer U, Hückelhoven R, Lee J, Scheel D (2015) A lectin S-domain receptor kinase mediates lipopolysaccharide sensing in Arabidopsis thaliana. Nat Immunol 16:426–433
9. Kutschera A, Dawid C, Gisch N, Schmid C, Raasch L, Gerster T, Schäffer M, Smakowska-Luzan E, Belkhadir Y, Vlot AC, Chandler CE, Schellenberger R, Schwudke D, Ernst RK, Dorey S, Hückelhoven R, Hofmann T, Ranf S (2019) Bacterial medium-chain 3-hydroxy fatty acid metabolites trigger immunity in Arabidopsis plants. Science 364:178–181
10. Schellenberger R, Crouzet J, Nickzad A, Shu LJ, Kutschera A, Gerster T, Borie N, Dawid C, Cloutier M, Villaume S, Dhondt-Cordelier S, Hubert J, Cordelier S, Mazeyrat-Gourbeyre F, Schmid C, Ongena M, Renault JH, Haudrechy A, Hofmann T, Baillieul F, Clement C, Zipfel C, Gauthier C, Deziel E, Ranf S, Dorey S (2021) Bacterial rhamnolipids and their 3-hydroxyalkanoate precursors activate Arabidopsis innate immunity through two independent mechanisms. Proc Natl Acad Sci U S A 118:e2101366118
11. Bigeard J, Colcombet J, Hirt H (2015) Signaling mechanisms in pattern-triggered immunity (PTI). Mol Plant 8:521–539
12. Sandoval PJ, Santiago J (2020) In vitro analytical approaches to study plant ligand-receptor interactions1. Plant Physiol 182:1697–1712
13. Xiao Y, Stegmann M, Han Z, DeFalco TA, Parys K, Xu L, Belkhadir Y, Zipfel C, Chai J (2019) Mechanisms of RALF peptide perception by a heterotypic receptor complex. Nature 572:270–274
14. Kato H, Nemoto K, Shimizu M, Abe A, Asai S, Ishihama N, Matsuoka S, Daimon T, Ojika M, Kawakita K, Onai K, Shirasu K, Yoshida M, Ishiura M, Takemoto D, Takano Y, Terauchi R (2022) Recognition of pathogen-derived sphingolipids in Arabidopsis. Science 376: 857–860
15. Liu T, Liu Z, Song C, Hu Y, Han Z, She J, Fan F, Wang J, Jin C, Chang J, Zhou J-M, Chai J (2012) Chitin-induced dimerization activates a plant immune receptor. Science 336:1160–1164
16. Shu L-J, Schäffer M, Eschrig S, Ranf S (2021) Low cost, medium throughput depletion-binding assay for screening S-domain-receptor ligand interactions using in planta protein expression. bioRxiv. https://doi.org/10.1101/2021.06.16.448648
17. Rosano GL, Ceccarelli EA (2014) Recombinant protein expression in Escherichia coli: advances and challenges. Front Microbiol 5: 172
18. van Oers MM, Pijlman GP, Vlak JM (2015) Thirty years of baculovirus–insect cell protein expression: from dark horse to mainstream technology. J Gen Virol 96:6–23
19. Schillberg S, Raven N, Spiegel H, Rasche S, Buntru M (2019) Critical analysis of the commercial potential of plants for the production of recombinant proteins. Front Plant Sci 10: 720
20. Martin R, Liu F, Staskawicz B (2022) Isolation of protein complexes from tobacco leaves by a two-step tandem affinity purification. Curr Protoc 2:e572
21. Lawson AW, Macha A, Neumann U, Gunkel M, Chai J, Behrmann E, Schulze-Lefert P (2025) Purifying recombinant proteins from *Nicotiana benthamiana* for structural studies. Nat Protoc (2025). https://doi.org/10.1038/s41596-025-01249-2
22. Eschrig S, Schäffer M, Shu L-J, Illig T, Eibel S, Fernandez A, Ranf S (2024) LORE receptor homomerization is required for 3-hydroxydecanoic acid-induced immune signaling and determines the natural variation of immunosensitivity within the Arabidopsis genus. New Phytol 242:2163–2179
23. Ranf S, Grimmer J, Poschl Y, Pecher P, Chinchilla D, Scheel D, Lee J (2012) Defense-related calcium signaling mutants uncovered via a quantitative high-throughput screen in *Arabidopsis thaliana*. Mol Plant 5: 115–130
24. Kaseniit KE, Katz N, Kolber NS, Call CC, Wengier DL, Cody WB, Sattely ES, Gao XJ (2023) Modular, programmable RNA sensing using ADAR editing in living cells. Nat Biotechnol 41(4):482–487
25. Lawrenson T, Youles M, Chhetry M, Clarke M, Harwood W, Hundleby P (2023) Efficient targeted mutagenesis in Brassica crops using CRISPR/Cas systems. In: Yang B, Harwood W, Que Q (eds) Plant genome engineering: methods and protocols. Springer, New York, pp 253–271
26. Jay F, Brioudes F, Voinnet O (2023) A contemporary reassessment of the enhanced transient expression system based on the tombusviral silencing suppressor protein P19. Plant J 113:186–204
27. Robert S, Khalf M, Goulet M-C, D'Aoust M-A, Sainsbury F, Michaud D (2013) Protection of recombinant mammalian antibodies from development-dependent proteolysis in

leaves of *Nicotiana benthamiana*. PLoS One 8: e70203

28. Alonso Villela SM, Kraïem H, Bouhaouala-Zahar B, Bideaux C, Aceves Lara CA, Fillaudeau L (2020) A protocol for recombinant protein quantification by densitometry. MicrobiologyOpen 9(6):1175–1182
29. Kwanyuen P, Wilson RF (2000) Optimization of coomassie staining for quantitative densitometry of soybean storage proteins in gradient gel electrophoresis. J Amer Oil Chem Soc 77: 1251–1254
30. Shimomura O, Kishi Y, Inouye S (1993) The relative rate of aequorin regeneration from apoaequorin and coelenterazine analogues. Biochem J 296:549–551
31. Inouye S, Shimomura O (1997) The use of Renilla luciferase, Oplophorus luciferase, and apoaequorin as bioluminescent reporter protein in the presence of coelenterazine analogues as substrate. Biochem Biophys Res Commun 233:349–353

Chapter 4

Plate Reader-Based Quantification of Cellular Calcium Dynamics in Immune Responses Using the Genetically Encoded Calcium Indicator GCaMP

Hyunsuh Lee, Robin Hon Ming Goh, Wolfgang Moeder, and Keiko Yoshioka

Abstract

Calcium ions (Ca^{2+}) serve as crucial secondary messengers in numerous signaling cascades in plants, similar to their role in animals. One of the earliest responses in the activation of immune responses is a rapid, transient increase in cytosolic Ca^{2+} ($[Ca^{2+}]_{cyt}$). Measuring the spatial and temporal changes in $[Ca^{2+}]_{cyt}$ is essential for understanding the details of immune responses to various pathogens and pests. We present a straightforward assay for measuring $[Ca^{2+}]_{cyt}$ using the genetically encoded Ca^{2+} indicator, GCaMP, and a fluorescence microplate reader. This assay allows for the assessment of changes in $[Ca^{2+}]_{cyt}$ with high temporal resolution following various elicitor treatments. Additionally, we describe a method to analyze $[Ca^{2+}]_{cyt}$ using a dexamethasone-inducible effector expression system (proDEX::AvrRpm1). Due to the simplicity of these methods, they are also suitable for high-throughput screening.

Key words Calcium signal, Plate reader, Genetically encoded calcium indicator, GCaMP, Immune response, Elicitors, Effectors, PTI, ETI

1 Introduction

Like in animals, Ca^{2+} acts as a secondary messenger in many signaling pathways in plants. Transient changes in cytosolic Ca^{2+} levels ($[Ca^{2+}]_{cyt}$) are one of the first detectable responses to many external stresses, including the interaction with pathogenic organisms [1]. The basis of Ca^{2+} signaling is a concentration gradient between the cytosol (100–200 nM) and the apoplast (or internal storage spaces), where $[Ca^{2+}]$ can be in the millimolar range [2]. With this steep concentration gradient, the opening of membrane-localized Ca^{2+}-conducting ion channels leads to rapid increases in $[Ca^{2+}]_{cyt}$.

Tsutomu Kawasaki and Hirofumi Yoshioka (eds.), *Plant Immunity: Methods and Protocols*, Methods in Molecular Biology, vol. 3012, https://doi.org/10.1007/978-1-0716-5138-4_4,

These changes can be sensed by a suite of Ca^{2+}-binding sensor proteins, which trigger appropriate downstream signaling cascades.

In plant immune responses, an increase in $[Ca^{2+}]_{cyt}$ is also believed to be one of the earliest responses of plant cells when they encounter pathogens or pathogen-derived molecules. In a first layer of immunity, plants recognize pathogen-associated molecular patterns (PAMPs) or plant-derived endogenous danger-associated molecular patterns (DAMPs) by plasma membrane-localized pattern recognition receptors (PRRs), which trigger pattern-triggered immunity (PTI) [1]. Plants have also evolved a second layer of defense, termed effector-triggered immunity (ETI), mediated by nucleotide-binding domain leucine-rich repeat proteins (NBS-LRRs or NLRs), which recognize either pathogen effectors directly or the cellular perturbations caused by these effectors. ETI often leads to localized cell death, known as the hypersensitive response (HR) [3].

PTI and ETI trigger a series of immune responses, including a reactive oxygen species (ROS) burst, the activation of mitogen-activated protein kinases (MAPKs), transcriptional reprogramming, induction of programmed cell death, and changes in phytohormone levels. In these immune responses, Ca^{2+} plays a significant role [4]. For example, NADPH oxidases, which play a key role in producing extracellular ROS, contain Ca^{2+}-binding EF-hand motifs and are therefore regulated by Ca^{2+} [4]. Recent data suggest that some pathogen effector receptors can function as Ca^{2+} channels to induce hypersensitive cell death [5].

To understand immune responses, it is crucial to monitor $[Ca^{2+}]_{cyt}$ changes at high spatial and temporal resolution. Initially, Ca^{2+}-sensitive fluorescent dyes such as Fura-2 or Indo-1 were utilized [6]; however, their use was hampered because the dye has to be taken up by the cell [7]. Aequorin-expressing transgenic plants were an improvement, but the aequorin cofactor coelenterazine still had to be supplied exogenously [8].

In the last decade, fluorescence-based genetically encoded Ca^{2+} indicators (GECIs) have dramatically improved the study of Ca^{2+} signaling, as they allow in vivo observation of dynamic Ca^{2+} fluxes without the need to apply a substrate [2, 7, 9]. Here, we use GCaMP, in which a permuted GFP is fused to the Ca^{2+}-binding protein calmodulin (CaM) and an M13 peptide, which binds to CaM in a Ca^{2+}-dependent manner. Ca^{2+} binding leads to GFP reconstitution and fluorescence emission, which can then be measured. This fluorescence is intensity-based and can be measured by a fluorescence microscope to obtain spatial information [10, 11]. On the other hand, for obtaining quantitative, temporal information at a high resolution, the usage of a plate reader can be a powerful tool [12]. Here, we describe a simple GCaMP3-based method to monitor $[Ca^{2+}]_{cyt}$ changes during PTI or ETI in

Arabidopsis thaliana leaf disks or young seedlings using a plate reader for 96-well or 24-well plates, respectively. This method can be applied to other plant species and can be combined with *Agrobacterium*-mediated transient expression or Virus-Induced Gene Silencing (VIGS) of genes-of-interest in *Nicotiana benthamiana* [12].

2 Materials

2.1 Materials for Both Arabidopsis Seedling and Mature Plant (Leaf Disk Methods)

1. *Arabidopsis thaliana* carrying GCaMP (*see* **Note 1**).
2. Growth chamber: We use a 9-h photoperiod, day/night temperature regime of 22 °C/20 °C to get plants that form large leaves before they start to bolt.
3. Fluorescence microscope with a GFP filter (*see* **Note 2**).
4. Microplate reader capable of measuring fluorescence at GCaMP3's excitation and emission ranges (GCaMP3 excitation peak = 488 nm, emission peak = 515 nm).
5. Immune elicitor of choice. Here, for PTI, we use two commonly used PAMPs, bacterial flagellin (flg22) and chitin. For ETI, we describe the utilization of an inducible pathogen effector expression system.
   (a) *flg22*: Prepare small aliquots of 100 μM stock solution in distilled water. Store the stock solution in −20 °C. Do not use the stock solution past three freeze-thaw cycles, as peptide can degrade. Before the experiment, prepare the working solution by diluting the stock solution to a desired volume and concentration. Final concentration of 100 nM to 1 μM gives a robust response.
   (b) *Chitin*:
      (i) Chitin oligosaccharides (e.g., TCI America, C2762): Dissolve chitin powder in distilled water to prepare a 1% (w/v) stock solution. Autoclave. Store the stock solution in dark, at 4 °C. Before the experiment, prepare the working solution by diluting the stock solution to a desired volume and concentration. Final concentration of 0.1% gives a robust response.
      (ii) Specific chitin oligomer (GN6–GN8): Before the experiment, dissolve the chitin powder in cold distilled water (*see* **Note 3**) to a desired volume and concentration. Final concentration of 100 μM to 1 mM should give a robust response (*see* **Note 4**).
   (c) *Effector*: Dexamethasone (DEX)-inducible pathogen effector using a stable transgenic Arabidopsis line carrying proDEX::AvrRpm1 [13].

2.2 Additional Materials for the Arabidopsis Seedling Method

1. Growth medium: ½ Murashige and Skoog (MS) liquid medium with 1% sucrose, pH 5.8, autoclaved.
2. Sterile 150 × 15 mm petri dish.
3. Micropore tape.
4. Black, clear foil-bottom 24-well plates (e.g., Sarstedt, lumox® multiwell, 94.6000.014).
5. Orbital shaker.

2.3 Additional Materials for the Mature Arabidopsis Plant (Leaf Disk) Method

1. Black, flat-bottom 96-well plates (e.g., Greiner, Cellstar®, 655079).
2. 4 mm biopsy punch (e.g., Integra, Miltex®, 3334) (*see* **Note 5**).
3. Multichannel pipette, reagent reservoir.

2.4 Additional Materials for Dexamethasone-Inducible Lines

1. Dexamethasone: Prepare 10 mM stock solution in DMSO. Store the stock solution in −20 °C. Before the experiment, prepare the working solution by diluting the stock solution with distilled water to a final concentration of 100 μM.

3 Methods

3.1 Arabidopsis Seedling Method

1. Grow Arabidopsis seedlings on a petri dish (sealed with micropore tape) in ½ MS liquid medium supplemented with 1% sucrose for about 9 days in the growth chamber (*see* **Note 6**). A minimum of 27 seedlings per genotype/treatment (9 seedlings per well, 3 wells) are needed for the assay.
2. On the day of the experiment, fill each well of the 24-well plate with 450 μL of ½ MS liquid medium supplemented with 1% sucrose. Use a minimum of 3 wells per genotype/treatment (4 or more wells per genotype/treatment are recommended). Remember to include wells for the no treatment (negative) control.
3. Using forceps, select and pool nine similarly sized GCaMP3-positive seedlings from the petri dish in which they were grown. Remove the group of seedlings from the petri dish and place them on a double-layered sheet of paper towel to remove excess media. Then, transfer the group of nine seedlings into a single well of the 24-well plate (containing 450 μL of media). Using forceps, gently spread the seedlings evenly within each well to minimize overlapping seedling at the bottom. Repeat until you have the desired number of wells for each genotype/treatment.
4. Place the lid on the 24-well plate. Incubate the plate on an orbital shaker at minimal speed for 2 h at room temperature.

The incubation allows the physical stress caused by the seedling transfer to subside.

5. Just before the end of the 2 h incubation:
   (a) Prepare the elicitor solutions for the microplate reader assay. Elicitors are prepared at 10× the final concentration and diluted to 1× final concentration upon addition in the well (seedlings incubating in 450 μL of media). Prepare 50 μL of 10× elicitor solution per well (*see* **Note 7**). Use distilled water without elicitor as the no treatment (negative) control.
   (b) Turn on and set up the microplate reader for a fluorescence measurement assay. Below are the settings we use on the Tecan Infinite M1000 Pro.
      (i) Plate definition: Select appropriate plate type
      (ii) Duration: 1 h
      (iii) Wavelength (*see* **Note 8**):
         1. Excitation: 470 nm, bandwidth 20 nm
         2. Emission: 525 nm, bandwidth 20 nm
      (iv) Mode: Top reading
      (v) Gain: Optimal (100% RFU)
      (vi) Flashes: Mode 1 [400 Hz], 50
         1. Settle time: 10 ms
      (vii) Z-position: 20400 μm
      (viii) TRF setting:
         1. Lag time: 0 μs
         2. Integration time: 20 μs
6. At the end of the 2 h incubation, remove the plate lid. Working quickly, add elicitor solution to each well (50 μL per well). Then, start fluorescence measurement immediately.
7. Fluorescence intensity will quickly increase, peaking at 10 min post flg22 or chitin treatment. Following the peak, fluorescence intensity will gradually decrease, reaching baseline level by 60 min post-treatment (Fig. 1).
8. Following the completion of the run, normalize the absolute fluorescence value of each replicate (i.e., well) across all time-points to its own final resting point fluorescence value (F/F_{final}). Then, take the average of the replicates, and graph it against time.

3.2 Mature Arabidopsis Plant (Leaf Disk) Method

1. Grow one GCaMP3 Arabidopsis plant per pot in the growth chamber. A minimum of 3 plants per genotype are needed for the assay (4 or more plants are recommended).

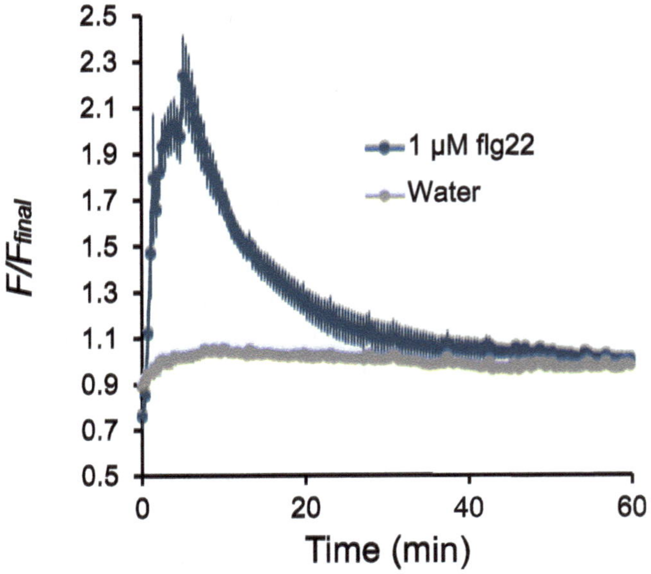

Fig. 1 Normalized $[Ca^{2+}]_{cyt}$ levels in 9-day-old Col-WT GCaMP3 seedlings following 1 μM flg22 treatment. Error bars show ± SE ($n = 3$, each well with 9 seedlings)

2. 4- to 5-week-old plants with fully expanded leaves (prior to bolting) are used for the assay. Approximately a week before the microplate reader assay, use a fluorescence microscope with a GFP filter to confirm GCaMP expression (*see* **Note 2**).
3. On the day of the experiment, select healthy-looking GCaMP3-expressing plants.
4. Using a multichannel pipette, fill each well of the 96-well plate with 100 μL of distilled water.
5. Punch leaf disks using a biopsy punch. Leaf disks can be taken from either side of the mid-vein. Avoid punching major veins. Select intermediate-sized leaves that look similar within and across plants. Be careful not to damage the disks prior to elicitor treatment. A minimum of three repeats (disks) can be used. However, more repeats are highly recommended. Adjust according to the number of lines/treatments of specific experiments.
6. Place one leaf disk per well (containing 100 μL of distilled water), abaxial side facing up. The leaf disk should float on top of the water.
7. Once finished, place the lid on the 96-well plate. Incubate the plate on the bench for 2–4 h at room temperature.
8. Just before the end of the incubation:
   (a) Prepare the elicitor solutions for the microplate reader assay. Elicitors are prepared at 2× the final concentration

and diluted to 1× final concentration upon addition in the well (leaf disks incubating in 100 μL of distilled water). Prepare 100 μL of 2× elicitor solution per well. Use distilled water without elicitor as the no treatment (negative) control.

(b) Turn on and set up the microplate reader for a fluorescence measurement assay. Below are the settings we use on the Tecan Infinite M1000 Pro.

(i) Plate definition: Select appropriate plate type.

(ii) Duration: 1 h

(iii) Wavelength (*see* **Note 8**):

1. Excitation: 470 nm, bandwidth 20 nm
2. Emission: 525 nm, bandwidth 20 nm

(iv) Mode: Top reading

(v) Gain: Manual, 90

(vi) Flashes: Mode 1 [400 Hz], 56

1. Settle time: 0 ms

(vii) Z-position: 20000 μm

(viii) TRF setting:

1. Lag time: 0 μs
2. Integration time: 20 μs

9. At the end of the 2–4 h incubation, remove the plate lid. Working quickly, add elicitor solution to each well (100 μL per well) using a multichannel pipette. Then, start fluorescence measurement immediately.
10. Following steps are identical to **steps 7**–**8** of seedling method.

Figure 2 shows $[Ca^{2+}]_{cyt}$ levels in leaf disks from GCaMP3 plants after treatment with flg22 and chitin oligosaccharide.

3.3 Using Dexamethasone (proDEX::AvrRpm1) for ETI Analysis (Leaf Disk)

Cross parental lines carrying GCaMP3 in WT or desired mutant background with a line carrying an inducible effector. Here, the *Pseudomonas syringe* effector AvrRpm1 under the control of a dexamethasone (DEX)-inducible promoter was used [13].

Follow the same protocol as above (leaf disk method **step 1**–7), then continue with the following modifications:

8. Prepare 100 μM DEX solution for the microplate reader assay. The DEX solution is prepared at 2× concentration and diluted to a final concentration of 50 μM upon addition to each well (leaf disks incubating in 100 μL of distilled water). Prepare 100 μL DEX solution per well. If a mock treatment is desired, use distilled water with an equivalent volume of DMSO.

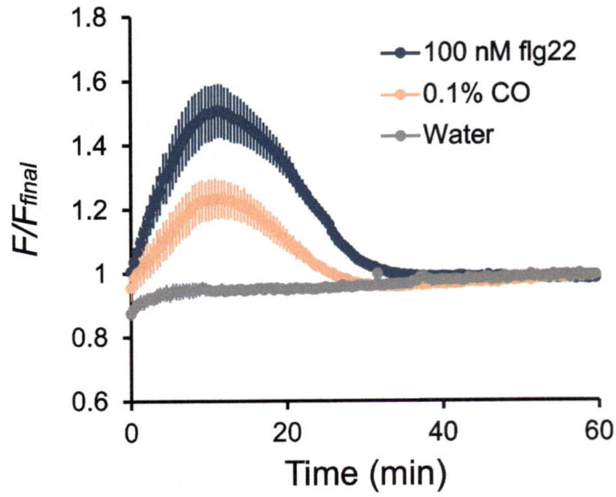

Fig. 2 Normalized $[Ca^{2+}]_{cyt}$ levels in leaf disks from 5-week-old Col-WT GCaMP3 plants following 100 nM flg22 or 0.1% chitin oligosaccharide (CO) treatment. Error bars show $\pm$ SE ($n = 10$)

9. At the end of the 2–4 h incubation, remove the plate lid. Working quickly, add DEX solution to each well (100 μL per well) using a multichannel pipette. Then, start fluorescence measurement immediately.
10. Fluorescence (for AvrRpm1) increases at 1 h, with a peak at 1.5–2 h post DEX treatment, and remains elevated.
11. Following the completion of the run, normalize the absolute fluorescence value of each replicate (i.e., well) across all time-points to its own initial ($t = 0$) fluorescence value (F/F_0). Then, take the average of the replicates, and graph it against time (*see* **Note 9**) (Fig. 3).

4 Notes

1. We use GCaMP3 plants [11], but the assay can also be used for other GCaMP versions.
2. Sometimes, within a plant line, some variation in the basal GCaMP3 expression level may occur due to the variation in the copy number or gene silencing. To avoid using plants with very low or no basal GCaMP3 expression level, use a fluorescence microscope with a GFP filter to confirm GCaMP3 expression prior to the microplate reader assay. Leaves from GCaMP3 expressing plants will appear bright under the fluorescence microscope compared to control leaves.
3. Larger chitin oligomers (longer chains) have lower solubility in water. Using cold water can help solubilize the chitin powder.

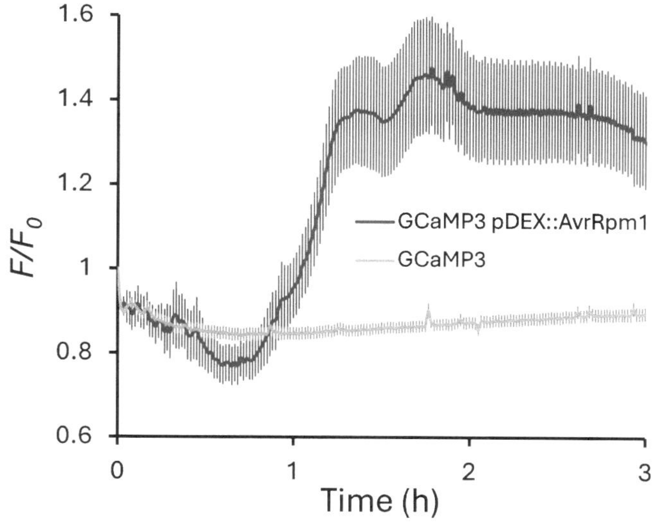

Fig. 3 Normalized $[Ca^{2+}]_{cyt}$ levels in leaf disks from 5-week-old Col-WT GCaMP3 and pDEX::AvrRpm1 GCaMP3 plants following treatment with 50 μM dexamethasone. Error bars show ± SE ($n = 9$)

Make sure that the chitin solution reaches room temperature before using it as an elicitor, to avoid cold shocking the plants.

4. The optimal concentration of specific chitin oligomers should be tested and adjusted accordingly, as the efficacy may vary depending on the size of the oligomer (GN8 > GN7 > GN6) and the purity/quality of the preparation. Different plant species and tissue types may also display different responsiveness. GN8 may give a response at a concentration of 1 μM.
5. If using genotypes with a morphological phenotype (e.g., small leaves) that makes it difficult to punch 4 mm leaf disks, use a 3 mm biopsy punch instead. In this case, take 3 mm leaf disks from all genotypes used in the assay.
6. If seedling growth appears inconsistent (e.g., varying in size) in liquid medium, germinate seeds on ½ MS agar media first, then transfer seedlings to liquid medium.
7. This elicitor solution (50 μL) to incubation media (450 μL) volume ratio ensures that the elicitor solution volume is large enough to allow for efficient diffusion in the well, but also not too large to cause unwanted osmotic changes.
8. This wavelength setting maximizes signal output by staying close to GCaMP3's excitation and emission peaks (excitation peak = 488 nm, emission peak = 515 nm), while minimizing excitation-emission overlap. Check your plate reader model for appropriate filters.
9. Depending on the type of treatment, the data normalization can be done against the starting (F/F_0) or the end point

fluorescence value (F/F_{final}). After PAMP treatment, we use the endpoint (F/F_{final}) for normalization, as the starting fluorescence values often show some variability. After induction of AvrRpm1, $[Ca^{2+}]_{cyt}$ remains elevated for a prolonged time; thus, here we use F/F_0.

Acknowledgments

We thank Dr. Hanae Kaku and Shingo Maruyama at Meiji University for their tireless guidance on chitin analysis. We also appreciate the technical assistance provided by the Centre for the Analysis of Genome Evolution and Function (CAGEF) at the University of Toronto. This work was supported by the NSERC Discovery Grant to KY (RGPIN-2019-05832).

References

1. Xu G, Moeder W, Yoshioka K, Shan L (2022) A tale of many families: calcium channels in plant immunity. Plant Cell 34:1551–1567
2. Grenzi M, Resentini F, Vanneste S, Zottini M, Bassi A, Costa A (2021) Illuminating the hidden world of calcium ions in plants with a universe of indicators. Plant Physiol 187:550–571
3. Jones JDG, Staskawicz BJ, Dangl JL (2024) The plant immune system: from discovery to deployment. Cell 187:2095–2116
4. Köster P, DeFalco TA, Zipfel C (2022) Ca^{2+} signals in plant immunity. EMBO J 41:e110741
5. Förderer A, Yu D, Li E, Chai J (2022) Resistosomes at the interface of pathogens and plants. Curr Opin Plant Biol 67:102212
6. Grynkiewicz G, Poenie M, Tsien RY (1985) A new generation of Ca2+ indicators with greatly improved fluorescence properties. J Biol Chem 260:3440–3450
7. Tian L, Andrew Hires S, Looger LL (2012) Imaging neuronal activity with genetically encoded calcium indicators. Cold Spring Harb Protoc 7:647–656
8. Knight H, Knight MR (1995) Recombinant aequorin methods for intracellular calcium measurement in plants. Methods Cell Biol 49:201–216
9. Toyota M, Betsuyaku S (2022) In vivo imaging enables understanding of seamless plant defense responses to wounding and pathogen attack. Plant Cell Physiol 63:1391–1404
10. Nagai T, Yamada S, Tominaga T, Ichikawa M, Miyawaki A (2004) Expanded dynamic range of fluorescent indicators for Ca(2+) by circularly permuted yellow fluorescent proteins. Proc Natl Acad Sci USA 101:10554–10559
11. Toyota M, Spencer D, Sawai-Toyota S, Jiaqi W, Zhang T, Koo AJ, Howe GA, Gilroy S (2018) Glutamate triggers long-distance, calcium-based plant defense signaling. Science 361:1112–1115
12. DeFalco TA, Toyota M, Phan V, Karia P, Moeder W, Gilroy S, Yoshioka K (2017) Using GCaMP3 to study Ca^{2+} signaling in Nicotiana species. Plant Cell Physiol 58:1173–1184
13. Mackey D, Holt BF, Wiig A, Dangl JL (2002) RIN4 interacts with Pseudomonas syringae type III effector molecules and is required for RPM1-mediated resistance in Arabidopsis. Cell 108:743–754

Chapter 5

Visualization of Calcium Decoding via a FRET-Based Reporter for Calcium-Decoder Conformation

Anja Liese, Sarah Lederer, Bernadette Eichstädt, and Tina Romeis

Abstract

Calcium sensors as tools to visualize calcium dynamics have significantly advanced our understanding of calcium signaling, but visualization of calcium decoding remains a major challenge. Towards this end, we developed the CPKaleon reporter for monitoring conformational changes of the calcium sensor-decoder family of calcium-dependent protein kinases (CPK). It enables the real-time visualization of CPK activation and inactivation in response to specific treatments. Here, detailed protocols for the in vitro characterization and in vivo imaging of CPKaleon are presented.

Key words Calcium-dependent protein kinases (CPKs), CPK conformational sensor, Förster resonance energy transfer, Live-cell imaging, Visualization of calcium decoding, Biosensor

1 Introduction

Kinases, as central enzymes in signal transduction pathways, act as molecular switches that regulate numerous cellular processes. However, monitoring kinase activity in vivo with spatial and temporal resolution remains a major challenge. In-gel kinase assays, which provide a "snapshot" of in vivo kinase activity, are a commonly used technique but do not permit real-time monitoring of cellular kinase activation dynamics [1–4].

Fluorescent-based kinase activity biosensors offer an ideal solution to overcome these limitations. Most kinase activity sensors are designed as substrate phosphorylation sensors, in which a substrate peptide sequence is sandwiched between a fluorescent protein (FP) pair. In these substrate-based sensors, peptide phosphorylation leads to altered Förster resonance energy transfer (FRET) efficiency [5–8].

The multi-member gene family of calcium-dependent protein kinases (CPKs in *Arabidopsis thaliana*) encodes essential kinases in plant calcium signalling. The conserved CPK structure is composed

Tsutomu Kawasaki and Hirofumi Yoshioka (eds.), *Plant Immunity: Methods and Protocols*, Methods in Molecular Biology, vol. 3012,
https://doi.org/10.1007/978-1-0716-5138-4_5,

of four domains: a N-terminal variable domain, a kinase domain, a pseudo-substrate segment (PS) and a calcium-binding calmodulin-like domain (CLD) [9]. Calcium decoding by CPKs can be broken down into three steps: (i) calcium binding to the CLD (sensing), which triggers (ii) a conformational change (required for CPK activation), and (iii) its relay into targeted protein phosphorylation (CPK activity). Because different CPK isoforms often are capable to phosphorylate identical target proteins, substrate-based FRET reporters may lack isoform-specific resolution. The calmodulin-based calcium sensor Cameleon has demonstrated that changes in protein conformation—triggered by calcium binding and substrate peptide interaction—can be monitored through changes in FRET efficiency [10]. Inspired by the Cameleon design, the CPK biosensor CPKaleon was developed to monitor calcium-dependent conformational changes within a CPK molecule itself (Fig. 1b) [11].

The CPK activity sensor CPKaleon is a Förster resonance energy transfer (FRET)-based biosensor. CPKaleon has a modular structure consisting of four modules: CPKpart_1, FRET donor, CPKpart_2 and the FRET acceptor (Fig. 1a). The N-terminal variable domain (V) and the kinase domain are part of CPKaleon sensor sequence N-terminal to the FRET donor (CPKpart_1). This first CPK-derived part of CPKaleon sequence encompasses the protein sequence from the first amino acid of a CPK sequence to the last amino acid upstream of the PS. The PS of CPKs consists of 31 amino acids and can be identified in CPK amino acid sequences based on sequence similarity. An amino acid WebLogo representing the 31 amino acids of the PS is shown in Fig. 1a [12]. The fluorescent proteins mTurquoise or eCFP are used as FRET donors, while cpVenus173 serves as FRET acceptor [13–15]. The FRET donor and acceptor fluorophores are interconnected by the PS and the main part of the CLD (CPKpart_2). The CLD is highly conserved, containing four calcium-binding EF-hand motifs. Note that several amino acids C-terminal to the fourth EF-hand exist in some isoforms, and hence are often designated as being a distinct "C-terminal variable domain" [16]. This short C-terminal amino acid sequence varies in length and amino acid composition between isoforms and is not included in CPKaleon constructs. The PROSITE entry PS50222 defines the EF-hand helix–loop–helix structure with a highly conserved calcium-binding loop [17, 18]. Based on PS50222, the fourth EF-hand can be identified defining the C-terminal end of CPKpart_2. The four different CPKaleon modules are connected via short linker sequences (2–3 amino acids). These linker sequences contain restriction enzyme recognition sequences compatible with classical cloning or serving as appropriate sequence overhangs for Golden Gate cloning [19] (Fig. 1a). The first generation of CPKaleon sensors has been designed as kinase-deficient variants to exclude an interference of enzyme auto-phosphorylation with the calcium-induced conformational

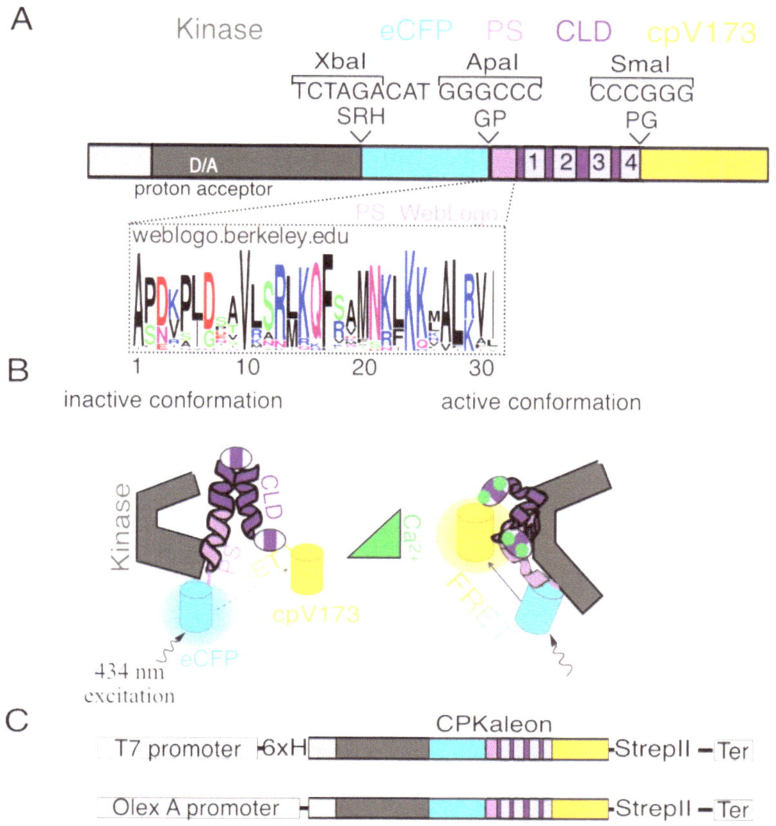

Fig. 1 Design and mechanism of CPKaleon. (**a**) CPKaleon uses eCFP as the donor FP and cpVenus173 as the acceptor FP, with the PS and CLD (CPKpart_2) in between. The WebLogo representation shows the PS amino acid sequence alignment of the *A. thaliana* CPK protein family. Six to nine nucleotides link the four different CPKaleon modules: CPKpart_1 (coding for the variable and kinase domain), the FRET donor, CPKpart_2 and the FRET acceptor. These linker sequences contain restriction sites, enabling classic restriction site cloning. (**b**) Calcium binding induces a conformational change that brings the FPs into closer proximity, leading to an increase in FRET efficiency. This conformational activation is reversed when calcium concentrations decrease. (**c**) The upper pictogram represents CPKaleon under a T7 RNA polymerase promoter, enabling inducible gene expression in *E. coli*. The lower pictogram shows the CPKaleon construct under the transcriptional control of an OlexA promoter, allowing inducible gene expression *in planta*. In this approach, a chimeric transcription factor, XVE, binds to the OlexA promoter after β-estradiol treatment. Inducible *in planta* gene expression systems are described in [26]. For fast and mild purification of the CPKaleon construct, a C-terminal StrepII-tag fusion was used. The CPKaleon fusion protein for *E. coli* expression has an additional N-terminal 6×HIS-tag

change. For this, a D/A amino acid substitution at the catalytically critical proton acceptor site for the substrate hydroxyl group has been chosen [20]. The proton acceptor is annotated in Prosite according to the UniRule PROSITE-ProRule: PRU00159 and PRU10027 [18].

To first investigate calcium responsiveness of the CPKaleon reporter with recombinant enzymes in vitro, a T7 RNA polymerase IPTG-inducible system was used to ensure high-level expression of CPKaleon in *Escherichia coli* [21] (Fig 1c). Purification was performed using an N-terminal 6×HIS-tag and a C-terminal StrepII tag, enabling selective enrichment of the full-length protein. To ensure accurate determination of CPKaleon calcium sensitivity, precautions regarding EGTA calcium buffering were taken and free calcium concentrations were calculated using the online tool "Maxchelator" [22].

For the use of the reporter in vivo, here, in *Arabidopsis thaliana* guard cells, we provide guidelines for sample preparation and image acquisition to monitor the CPK conformational activation and inactivation in a FRET-based manner. For this, the CPKaleon in planta expression construct is under the control of a β-estradiol-inducible promoter, allowing localized and time-restricted CPKaleon expression (Fig. 1c). This minimizes the risk of developmental alterations caused by disturbances in calcium homeostasis or signalling.

The design of the CPKaleon sensor described here is suitable for both in vitro and in vivo measurements [11]. Future CPK activity sensors may incorporate a variety of fluorescent protein FRET pairs, linkers of different lengths, kinase-active or -inactive variants and alternative promoters. This would enable researchers to select optimized features based on their experimental needs and the specific CPK isoform being analyzed.

In addition to the CPKaleon biosensor, we provide a protocol for imaging CPKaleon time series and simultaneous recording of calcium dynamics via an intensiometric calcium sensor. Dual-sensor plant material was generated by transforming the CPKaleon construct into the previously described *R-GECO1* transgenic line [23]. Guidelines for sample preparation, image acquisition and image analysis using Fiji and an R-script have been optimized for in vivo imaging in guard cells but can be easily adapted for use in other cell types.

2 Materials

Prepare all solutions using analytical grade reagents in 18 MΩ deionized water. Prepare and store all reagents at room temperature (unless indicated otherwise).

2.1 In Vitro Analyses of CPK-FRET

2.1.1 Recombinant CPKaleon Expression in E. coli

1. Lysogeny broth (LB) media: Dissolve 25 g LB in 1 L water. To prepare LB plates 1.5% (w/v) bacterial agar was added. Autoclave LB media for 20 min at 121 °C.
2. 10 x Terrific-Broth (TB) salts: Dissolve 23.1 g KH_2PO_4 and 125.4 g K_2HPO_4 in 1 L water. Autoclave for 20 min at 121 °C.
3. Terrific-Broth (TB) media: Dissolve 12 g tryptone, 24 g yeast extract and 5 g glycerol in 900 mL water. Autoclave for 20 min at 121 °C and let it cool down to room temperature before adding 100 mL 10 x TB salts.
4. 100 mM Isopropyl-β-d-thiogalactopyranoside (IPTG): Dissolve 1.19 g in 50 mL water. Filter-sterilise the solution using a 0.45 μm filter and store as 1 mL aliquots at −20 °C.
5. *E. coli* cells: Novagen BL21 D3 pLysS cell strain transformed with a plasmid coding for CPKaleon under the transcriptional control of a T7 RNA polymerase promoter (Fig. 1c). *E. coli* strain genotype: F^- *ompT* *hsdS*$_B$(r_B^- m_B^-) *gal dcm* (DE3) pLysS (Cam^R). A detailed protocol for the chemical transformation of *E. coli* is provided in [24].

2.1.2 Protein Purification from E. coli

1. 1 M Dithiothreitol (DTT) stock solution: Dissolve 15.42 g in 100 mL water. Filter-sterilise the solution using a 0.45 μm filter and store 1 mL aliquots at −20 °C. Avoid repeated freeze-thaw cycles.
2. Lysis buffer: 50 mM Hepes-KOH, pH 7.4, 300 mM NaCl, 1: 200 protease inhibitor cocktail for Histidine-tagged proteins, 1 mM DTT, 30 mM imidazole. Just before resuspending the *E. coli* cells, add 100 μL of 1 M DTT to 100 mL of the buffer solution.
3. Lyophilized lysozyme powder.
4. Bandelin SONOPULS ultrasonic homogenizers (e.g., HD 3100) with a titanium alloy sonotrode (e.g., MS 72) or a similar system for disrupting *E. coli* cells.
5. Wash buffer H: 50 mM Hepes-KOH, pH 7.4, 300 mM NaCl.
6. Wash buffer H containing 500 mM imidazole (*see* **Note 1**).
7. Merck Ni Sepharose™ 6 Fast Flow.
8. Bio-Rad empty gravity-flow polypropylene column.
9. Wash buffer S: 100 mM Tris–HCl, pH 8.0, 150 mM NaCl.
10. Elution buffer S: 100 mM Tris–HCl, pH 8.0, 300 mM NaCl, 10 mM $MgCl_2$, 2.5 mM β-desthiobiotin. Dissolve 5.36 mg β-desthiobiotin in 10 mL 100 mM Tris–HCl, pH 8.0, 300 mM NaCl, 10 mM $MgCl_2$. Store at 4 °C up to 1 week.
11. IBA Strep-Tactin® MacroPrep® resin.
12. To prepare a disposable column for Strep-Tactin® MacroPrep® resin, close the column using the bottom cap and fill it with

2 mL of wash buffer S. Float a porous disc on top of the liquid and push it to the bottom of the column using the blunt end of a Pasteur pipette. Open the column by removing the bottom cap and allow the liquid to flow through. Before the liquid is completely drained, close the column.

13. Roth Micro Dialysis Capsules QuixSepR.
14. Dialysis membrane with a 6000–8000 Da cutoff.
15. Dialysis buffer: 30 mM Tris–HCl, pH 7.4, 150 mM NaCl and 10 mM $MgCl_2$.
16. SDS-PAGE: A detailed protocol of SDS polyacrylamide gel preparation and gel electrophoresis is provided by Kurien and Scofield [25].
17. Bio-Rad protein assay dye working solution: Dilute 1 part protein assay dye reagent concentrate with 4 parts water.
18. Bovine serum albumin (BSA): Dissolve 1 mg BSA in 1 mL water. Dilute this BSA stock solution with water to obtain 0.5 mg/mL, 0.4 mg/mL, 0.3 mg/mL, 0.2 mg/mL, 0.1 mg/mL and 0.05 mg/mL BSA concentration standards.
19. GRP single-step protein gel stain "Der Blaue Jonas."
20. A microplate absorbance reader set to 595 nm.

2.1.3 Preparation of Calcium Buffers

Always calibrate your pH-meter before adjusting the pH according to the manufacturer's instructions (*see* **Note 2**).

1. 200 mM EGTA Stock solution: To dissolve EGTA in water, weigh 3.8 g of EGTA and add water to a final volume of 90 mL. Adjust the pH to 8.0 with NaOH while continuously mixing. Deprotonation of EGTA's four acidic groups increases with higher pH, enhancing its solubility in water. A pH below 8.0 may result in partial solubilization of EGTA. Finally, bring the solution to a total volume of 100 mL with water.
2. Zero-Ca^{2+}-buffer (*see* **Note 3**): 30 mM Tris–HCl, pH 7.4, 150 mM NaCl, 10 mM $MgCl_2$, 20 mM EGTA. Transfer 10 mL of a 200 mM EGTA solution into a 100 mL graduated cylinder. Add 6 mL of 2500 mM NaCl and 3 mL of 1 M Tris–HCl, pH 7.4, and 10 mL of 1 M $MgCl_2$ stock solution. Add water to a total volume of 80 mL. Divide the solution into two 40 mL portions. For one portion adjust the pH to 7.4. at 22 °C. Fill up to a volume of 50 mL with water and filter-sterilise this solution. Store this zero-calcium-buffer at 4–8 °C.
3. 20 mM Ca^{2+}-buffer: 30 mM Tris–HCl, pH 7.4, 150 mM NaCl, 10 mM $MgCl_2$, 20 mM $CaCl_2$, 20 mM EGTA. Add 1 mL of 1 M $CaCl_2$ to the second portion of the solution prepared in **step 2**. Adjust the pH with NaOH to pH 7.4 at 22 °C (*see* **Note 4**). Bring the volume to 50 mL with water. Filter-sterilise this solution and store at 4–8 °C.

2.1.4 In Vitro Measurement of CPK-FRET Conformational Change

1. A microplate fluorescence reader capable of fluorescence emission spectral scanning with an excitation of approximately 435 nm, such as TECAN Infinite M200 PRO plate reader.
2. PerkinElmer black OptiPlate 96 well.
3. Microsoft Excel or equivalent spreadsheets organized program for data computing with formulas and functions.
4. GraphPad Prism 5 or equivalent graphing and analysis program suitable for fitting biological data. GraphPad Prism is not available for free, and a license is required. A free 30-day trial can be requested on www.graphpad.com.

2.2 Live-Cell Fluorescence Imaging in Guard Cells

2.2.1 Plant Growth

1. Jiffy-7 pellets plant growth soil.
2. 3–4 weeks old transgenic *pOlexA:CPKaleon* (Fig. 1c), *pUBI:R-GECO1* [23] *Arabidopsis thaliana* plants.

2.2.2 Sample Preparation of Plant Material

1. IBIDI chambered coverslip with two wells for use in high-end microscopy.
2. Ullrich Medical Adhesive B liquid.
3. Scotch magic tape.
4. 40 mM 17-β-estradiol stock solution: Dissolve 10.9 mg 17-β-estradiol in 1 mL DMSO. Store at 4 °C in the dark for up to 1 week. Note that 17-β-estradiol is sensitive to light and temperature, therefore, avoid freeze-thaw cycles [26].
5. Plant imaging buffer with β-estradiol: 10 mM MES-Tris, pH 6.15, 5 mM KCl, 50 μM $CaCl_2$, 20 μM 17-β-estradiol and 0.05% DMSO. Dissolve 0.195 g MES in 80 mL water add 0.5 mL of 1 M KCl stock solution and 5 μL 1 mM $CaCl_2$ stock solution. Mix and adjust the pH with 100 mM Tris solution. Fill up to 100 mL with water. Filter-sterilise and store in the dark. Add 50 μL 40 mM 17-β-estradiol stock solution to 100 mL plant imaging buffer just before adding to the prepared epidermal peels.

2.2.3 In Vivo Imaging

1. Treatment stock solution: 5 μM flg22 in water or 1 mM ABA in 10% ethanol (*see* **Note 5**).
2. A Confocal Laser Scanning Microscope. All in vivo imaging experiments were performed on a Zeiss LSM 880 system with a 40× water immersion objective (LD C-APOCHROMAT, 40×/1.1 Korr UV-VIS M27, Zeiss).
3. For selecting an open and reporter expressing stoma by eye, a white light halogen lamp (e.g., Zeiss HAL100) and a metal halide fluorescence lamp (e.g., Zeiss HXP 120 V) were used.
4. For digital image acquisition, an Argon multiline laser (458 nm, 514 nm) and a DPSS laser (561 nm) were used.
5. The imaging system was controlled using Zen Black software.

2.2.4 Imaging Data Analyses

1. Fiji (Fiji Is Just ImageJ) is an image processing package based on ImageJ [27]. Fiji is an open-source software and freely available from https://imagej.net/Fiji/Downloads.
2. Microsoft Excel and GraphPad Prism 5 data analysis programs (*see* Subheading 2.1.4, **steps 3** and **4**).
3. R using R studio as graphical user interface (GUI). R is an open-source software environment for statistical analysis and graphics. R is freely available from R: The R Project for Statistical Computing [28] and RStudio from Posit https://posit.co/.

3 Methods

3.1 In Vitro Analyses of CPK-FRET

Glycerol stock of the *E. coli* BL21 (DE3) pLysS strain transformed with CPKaleon construct is stored at −80 °C.

3.1.1 Recombinant CPKaleon Expression in E. coli

1. Streak the corresponding glycerol stock cultures onto LB agar plates containing 50 μg/mL kanamycin and 34 μg/mL chloramphenicol, then incubate overnight (12–16 h) at 37 °C.
2. Prepare a preculture by picking a single colony and inoculating it into 20 mL of LB broth medium containing 50 μg/mL kanamycin and 34 μg/mL chloramphenicol. Incubate the culture in a 100 mL Erlenmeyer flask overnight at 120 rpm and 37 °C.
3. Dilute the preculture 1:100 to a final volume of 1 L of TB medium containing 50 μg/mL kanamycin and 34 μg/mL chloramphenicol. The starting optical density at 600 nm (OD600) for the main culture should be between 0.04 and 0.06. Adjust the volume of TB medium or preculture accordingly.
4. Grow the main culture in a 2 L baffled flask at 37 °C and 120 rpm until an OD600 of 0.5 to 0.8 is reached (*see* **Note 6**).
5. Add IPTG to the TB medium at a final concentration of 0.4 mM, then incubate at 120 rpm and 22 °C for 4 h.
6. Harvest the bacteria by centrifugation at 3000× *g* for 10 min at 4 °C, and freeze the resulting pellets at −20 °C.

3.1.2 Protein Purification from E. coli

Perform all protein purification steps at approximately 4 °C (unless indicated otherwise) to ensure the stability of the CPKaleon protein.

1. Resuspend the bacterial pellet in 40 mL of lysis buffer.
2. Add a spatula tip of lysozyme and incubate the mixture for 15 min.
3. Lyse the cells on ice via sonification for 20 min, with a cycle of 30 s of sonication followed by 20 s of rest.

4. Add Triton X-100 to a final concentration of 0.2% (v/v), incubate for 5 min and centrifuge at 20,000× *g* for 20 min.
5. Filter the supernatant through a 75-μm nylon mesh to remove any remaining cell debris from the lysate.
6. Transfer 600 μL of Ni Sepharose™ 6 Fast Flow (50% (w/v) suspension in 20% EtOH) to a 50 mL reaction vessel and wash with 1 mL water twice. Equilibrate the Ni Sepharose with 1 mL wash buffer H three times. To wash and equilibrate the matrix, gently shake and centrifuge at 500× *g* for 3 min and replace the supernatant by new solution.
7. Apply the cell lysate to the equilibrated Ni Sepharose matrix and incubate for 50 min on a rotating wheel at 10 rpm.
8. Centrifuge the Ni Sepharose-lysate mixture at 500× *g* for 5 min. Discard the supernatant, then add 2 mL of wash buffer H containing 30 mM imidazole to the matrix and centrifuge at 500× *g* for 5 min.
9. Discard the supernatant and add 2 mL of wash buffer H containing 30 mM imidazole. Resuspend the matrix by gently shaking and transfer the solution to an empty gravity-flow polypropylene column.
10. Wash twice with 2 mL of wash buffer H containing 40 mM imidazole.
11. Once the wash solution has nearly completely run through, close the column with a bottom cap and add 1 mL of room-temperature wash buffer H containing 500 mM imidazole. Incubate for 10 min at room temperature, then reopen the column and collect the eluate.
12. Mix the eluate in a 5 mL vessel with 3 mL of wash buffer S and 600 μL of Strep-Tactin MacroPrep (50% (w/v) suspension).
13. Incubate for 50 min on a rotating wheel at 10 rpm.
14. Apply the Strep-Tactin-bound CPKaleon to an empty prepared gravity-flow column.
15. Close the column with the bottom cap.
16. Once the Strep-Tactin matrix has completely settled on the column filter reopen the column.
17. Wash twice with 2 mL of wash buffer S.
18. Wash with 0.3 mL of elution buffer S.
19. Elute the recombinant Strep-tag II fusion protein three times with 0.5 mL of room-temperature elution buffer S. After adding 0.5 mL of elution buffer, close the column with a bottom cap and incubate for 10 min at room temperature. Reopen the column and collect every eluate fraction. Protein concentration of each fraction should be analyzed using the Bio-Rad protein assay.

20. Pipet 10 μL of 0.05–0.5 mg/mL BSA concentration standards and undiluted as well as 1:4 diluted elution fractions into microtiter plate wells.
21. Add 200 μL of 1x Bio-Rad protein assay dye working solution to each well.
22. Incubate at room temperature for 5 min.
23. Measure absorbance at 595 nm.
24. Analyze 10 μL samples of each elution fraction by SDS-PAGE.
25. Stain SDS-gel with separated protein probes in 50 mL single-step protein gel dye “Der Blaue Jonas” for at least 10 min. Staining intensity will increase up to 1 h.
26. The fraction with the highest protein concentration is transferred to a 1 mL quick-dialysis capsule sealed with a dialysis membrane with a 6000–8000 Da cutoff. Transfer the prepared quick-dialysis capsule to 1.5 L of dialysis buffer and dialyze for 4 h (approximately 45 min per 100 μL eluate). Continuously mix the dialysis solution during dialysis on a magnetic stirrer at 4 °C.
27. After dialysis measure protein concentration using the Bio-Rad protein assay.
28. Analyze 10 μL of dialyse protein by SDS-PAGE and “Der Blaue Jonas” staining.
29. Dilute the protein with dialysis buffer to a final concentration of 415 nM (*see* **Note 7**).

3.1.3 Preparation of Calcium Buffers

1. Prepare calcium buffers covering a range of calcium concentrations by performing reciprocal dilutions of a zero-Ca^{2+} buffer with a 20 mM-Ca^{2+} buffer.
2. To create a range of calcium-EGTA-buffers with increasing calcium concentration, the total calcium concentration is calculated using the equation:

$$V_{\text{replace}} = V_{\text{final}} \times (x - y) \div (z - y)$$

x = calcium concentration after dilution, y = calcium concentration before dilution, z = calcium in the high calcium buffer (20 mM), V_{final} = volume of each sample (1 mL).

3. The preparation of calcium-EGTA buffers is a stepwise process:
4. First step: For each buffer numbered 2–15 (see Table 1), add 1mL of zero-calcium-buffer to reaction tube. Replace 0.1 mL in each buffer with 20 mM-calcium-buffer. Buffer number 2 is complete and can be stored at 4 °C.
5. Second step: Replace for each buffer numbered 3–15, 0.167 mL with 20 mM-calcium-buffer. Buffer number 3 is complete and can be stored at 4 °C.

Table 1
Reciprocal dilutions used to create a range of calcium-EGTA buffers

	Calcium-buffer		Final concentration				
Buffer-number	Total Ca^{2+} [mM]	Total EGTA [mM]	Total Ca^{2+} [mM]	Total EGTA [mM]	Final volume mL	Volume to replace mL	Free Ca^{2+} [nM]
1	0	20	0	10	1	0.000	0
2	2	20	1	10	1	0.100	13
3	5	20	2.5	10	1	0.167	41
4	6.6	20	3.3	10	1	0.107	61
5	8	20	4	10	1	0.104	95
6	11	20	5.5	10	1	0.250	175
7	12	20	6	10	1	0.111	215
8	14	20	7	10	1	0.250	335
9	16	20	8	10	1	0.333	575
10	16.6	20	8.3	10	1	0.150	702
11	17	20	8.5	10	1	0.118	815
12	17.4	20	8.7	10	1	0.133	962
13	18	20	9	10	1	0.231	1200
14	19	20	9.5	10	1	0.500	2700
15	19.4	20	9.7	10	1	0.400	4500
16	20	20	10	10	1	1.000	38300

6. Third step: Replace for each buffer numbered with 4–15, 0.107 mL with 20 mM-calcium-buffer. Buffer number 4 is complete and can be stored at 4 °C. Continue replacing the indicated volumes as shown in Table 1 until reaching Buffer 15.
7. For buffer numbered 1 and 16 use directly the zero-calcium-buffer or the 20 mM-calcium-buffer.
8. Calculate the free Ca^{2+} concentrations indicated in Table 1 using the WEBMAXC Extended website (http://tinyurl.com/y48t33xq). Values for total calcium, total EGTA concentration (*see* **Note 8**), and calculated ionic strength (see below) are listed in Tables 1 and 2. For the calculations, set the pH to 7.4, and set the temperature to 22 °C. The mathematical basis for determining free Ca^{2+} concentrations with this program is described in detail in [22].
9. Calculate the ionic strength as a function of the concentration of all salts in the solution. It can be determined using the formula:

Table 2
Calculated ionic strength in the calcium-EGTA buffers. Ionic strength values for zero-Ca^{2+} buffer (A) and for calcium-EGTA buffers (B) are provided

A

Buffer	c [M]*	|z|	Ionic strength [N]
[Na^+] from NaCl	0.15	1	0.075
[Cl^-] from NaCL	0.15	1	0.075
[Mg^{2+}] from $MgCl_2$	0.01	2	0.02
[2*Cl^-] from $MgCl_2$	0.02	1	0.01
[$Tris^+$] from Tris	0.004	1	0.002
[EGTA2^-] from EGTA	0.01	2	0.02
Σ Ionic strength calculated without $CaCl_2$			0.202

B

[Ca^{2+}] from $CaCl_2$ [M]	|z|	Ionic strength [N]	[2*Cl^-] from $CaCl_2$ [M]	|z|	Ionic strength [N]	Σ Ionic strength Buffer [N]
0	2	0	0	1	0	0.202
0.001	2	0.002	0.002	1	0.001	0.205
0.0025	2	0.005	0.005	1	0.0025	0.210
0.0033	2	0.0066	0.0066	1	0.0033	0.212
0.004	2	0.008	0.008	1	0.004	0.214
0.0055	2	0.011	0.011	1	0.0055	0.219
0.006	2	0.012	0.012	1	0.006	0.220
0.007	2	0.014	0.014	1	0.007	0.223
0.008	2	0.016	0.016	1	0.008	0.226
0.0083	2	0.0166	0.0166	1	0.0083	0.227
0.0085	2	0.017	0.017	1	0.0085	0.228
0.0087	2	0.0174	0.0174	1	0.0087	0.228
0.009	2	0.018	0.018	1	0.009	0.229
0.0095	2	0.019	0.019	1	0.0095	0.231
0.0097	2	0.0194	0.0194	1	0.0097	0.231
0.01	2	0.02	0.02	1	0.01	0.232

$$I = \frac{1}{2} \times \sum c_i \times z_i^2,$$

where c_i is the molar concentration of ion, z_i is the charge of that ion and the sum is taken over all ions in the solution (*see* **Notes 9** and **10**). Calculated ionic strength values and the variables used for each salt are listed in Table 2.

3.1.4 In Vitro Measurement of CPK-FRET Conformational Change

1. Pipette 25 μL of calcium-EGTA-buffer into a 96-well plate. We recommend using a black 96-well OptiPlate because it exhibits lower autofluorescence compared to fluorescence-compatible 96-well plates from other providers.
2. Pipette 25 μL of the dialyzed CPKaleon protein solution into the wells.
3. Measure emission within the range of 470–600 nm using an excitation wavelength of 435 nm (bandwidth: 5 nm) in a fluorescence plate reader. Monitor emission in 2 nm steps with 10 flashes of 20 μs and 400 Hz using a TECAN Infinite M200 PRO plate reader. Data are exported in Excel.
4. Calculate FRET emission ratios by dividing the cpVenus173 emission maxima by the eCFP emission maxima. For eCFP, determine the maxima in the emission range of 470–490 nm, and for cpVenus173, in the emission range of 518–538 nm (Fig. 2).
5. Data are further processed in GraphPad Prism 5 (*see* **Note 11**). The FRET emission ratios (y-axis values) are plotted against the decadic logarithms of the calcium concentration (x-axis values) in an X-Y-chart (Fig. 2).
6. In GraphPad Prism 5, navigate to "Analyze" > "XY analyses" > "Nonlinear regression (curve fit)." Click "ok." In the Parameters: "Nonlinear regression" window, set the options "Sigmoidal dose-response variable slope" and confirm with "ok."
7. The program uses a four-parameter logistic equation as a dose-response model to fit the FRET emission ratios dependent on the logarithm of the increasing free Ca^{2+} concentrations (Fig. 2). The equation can be written as:

$$Y = \text{Bottom} + \frac{Top - \text{Bottom}}{1 + \left(\frac{10^{\text{LogEC}_{50}}}{10^{x}}\right)^{\text{HillSlope}}}$$

x = free [Ca^{2+}], y = FRET emission ratio.

8. Based on your data the program determines the best fit value for Bottom, Top, EC_{50} and HillSlope.

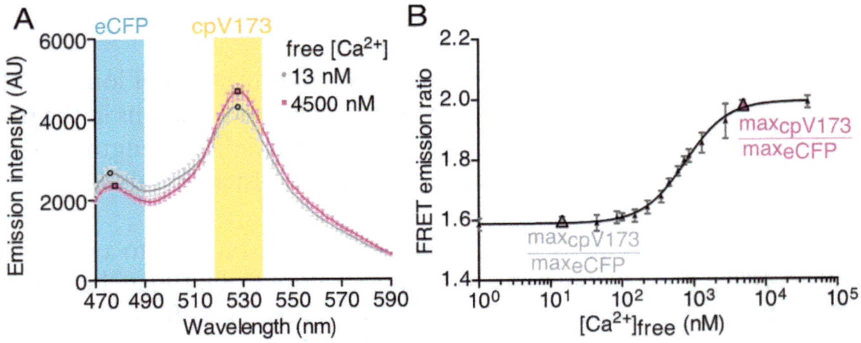

Fig. 2 In vitro analyses of CPKaleon calcium sensitivity. (**a**) CPKaleon emission spectra (470–590 nm, step size: 2 nm) at 435 nm excitation. (**b**) The FRET emission ratios plotted against the decadic logarithms of free $[Ca^{2+}]$

9. Use the best value obtained for Bottom to calculate ΔFRET as percentage change of emission ratios. ΔFRET is calculated using the equation:

$$\Delta\text{FRET}\left[_{\text{free}}Ca^{2+}\right]_{x\,\mu M} = \frac{\text{FRET ratio}\left[Ca^{2+}\right]_{x\,\mu M}}{\text{FRET ratio Bottom}} \times 100 - 100$$

10. Fit the percentage change of emission ratio ΔFRET using a four-parameter logistic equation as described in **step 6** (*see* **Note 12**).

3.2 Live-Cell Fluorescence Imaging in Guard Cells

Live-cell fluorescence imaging was performed in guard cells of epidermal peels.

3.2.1 Plant Growth

Cultivate plants for 2–3 weeks on Jiffy-7 soil using a cycle of 10 h light and 14 h dark at a temperature of 20–22 °C, 60% RH and 120 μE/m^2 light intensity.

3.2.2 Sample Preparation (Fig. 3)

1. Coat a two-well chambered coverslip with a thin layer of medical adhesive (Fig. 3b).
2. Let the volatile components evaporate for 10 min under the fume hood. Wash three times with 1 mL water.
3. Cut a small leaf part and gently press the abaxial part of the leaf to the medical adhesive layer (Fig. 3c).
4. Carefully tape the Scotch magic tape to the adaxial epidermis (Fig. 3d).
5. Peel the adaxial epidermis and mesophyll cell layer away by removing the magic tape. Repeat affixing and pulling away with magic tape, leaving a single-layer lower epidermis (Fig. 3e).

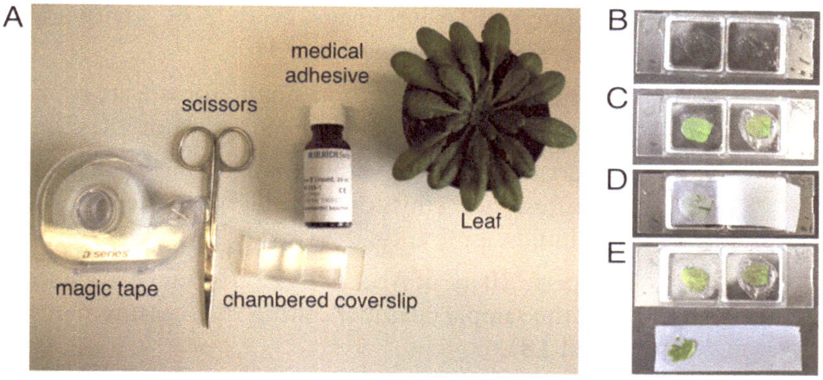

Fig. 3 Sample preparation for in vivo imaging. (**a**) Materials used for sample preparation. (**b**–**e**) Sample preparation steps. (**b**) Medical adhesive-coated two-well chambered. (**c**) Two-well chamber with fixed leaf. (**d**) Magic tape taped to the adaxial epidermis. (**e**) Prepared epidermal strip by removing most leaf layers with magic tape

6. Immerse the chamber coverslip in 1 mL plant imaging buffer with β-estradiol.
7. Transfer the prepared samples back in the plant growth chamber (conditions described under see Subheading 3.2.1) and incubate overnight for recovery.
8. Exchange the buffer to fresh plant imaging buffer with β-estradiol.
9. Incubate the samples under light at 22–24 °C for 2–3 h before imaging.

3.2.3 Imaging

1. Perform the imaging in bottom imaging mode on an inverse confocal laser scanning microscope with a 40× water immersion objective.
2. Focus on guard cells using bright-field illumination.
3. Select well-expressing, open guard cells in the YFP channel, with 514 nm excitation and an emission range between 525 and 560 nm for YFP (*see* **Note 13**).
4. To enhance signal intensity, open the pinhole to 599 μm (*see* **Note 14**).
5. For excitation of eCFP use the argon laser at 458 nm, and for R-GECO1, the DPSS laser (561 nm). For detection of eCFP emission, use a range from 465 to 505 nm, for cpVenus173 from 525 to 560 and for cpmApple in R-GECO1 from 580 to 611 nm. Set the detector gain and laser intensity percentage so that most of the detected pixels fall within the lower dynamic intensity range of your detector. This can be visualized in most image processing software using a histogram, where the absolute frequency of pixels (y-axis) is plotted against the dynamic intensity range of the detector (x-axis) (*see* **Note 15**).

6. Acquire a 16-bit image every 3–5 s with a frame size of 512 × 512 pixels (*see* **Note 16**).
7. If both guard cells from a single stomata are expressing, analyse them separately as individual regions of interest (ROIs).
8. Start a time series with a frame rate of 3–6 s for 45 min. The fluorescence emission should be stable for 3–5 min before stimulus application.
9. Pipette 20 μL of ABA or flg22 treatment stock solution directly to the sample chamber during the measurement (*see* **Notes 17** and **18**).

3.2.4 Imaging Data Analyses

Here, post-acquisition analyses of the time series focus on analysing fluorescence intensity dynamics over time. Manual image analysis in Fiji involves the following steps: (1) reducing image noise using a Gaussian blur filter, (2) defining a fluorescence intensity threshold, (3) measuring fluorescence intensities in selected regions of interest (ROIs) for each frame. Further processing of fluorescence intensity data in data analysis programs, here Excel and GraphPad, includes: (1) calculating fluorescence ratios for CPKaleon FRET donor-acceptor signals, (2) graphically representing intensity changes over time using plots. Synchronization analysis between CPKaleon fluorescence ratio changes and R-GECO1 fluorescence intensity changes is conducted using an adapted R script, phase_analysis, as previously described in [29].

1. Open the recorded time-series in Fiji "File" > "Open".
2. The "Bio-Formats Importer Options" window opens automatically. Select under the option "View Stack with" "Hyperstack" and for the option "Split into separate windows" "Split channels".
3. Perform **steps 4–11** for the R-GECO1 (GECO1 excitation/R-GECO1 emission), cpVenus-FRET (CFP excitation/YFP emission) and CFP (CFP excitation/CFP emission) image series.
4. Reduce image noise by blurring the images of the time series with a Gaussian function. Go to "Process" > "Filters" > "Gaussian Blur." Set the "Sigma Radius" to 1.
5. To convert the 16-bit grayscale image to a 32-bit grayscale image go to "Image" > "Type" > "32-bit".
6. Save the image stacks as TIFF file.
7. To achieve a better signal-to-noise ratio, define the imaging threshold by navigating to "Image" > "Adjust" > "Threshold". In the Threshold window, select "Dark background" and enable "Stack histogram". Define the selected pixel intensity range in the stack histogram so that all signal (highlighted in red) is included, while excluding any background (highlighted

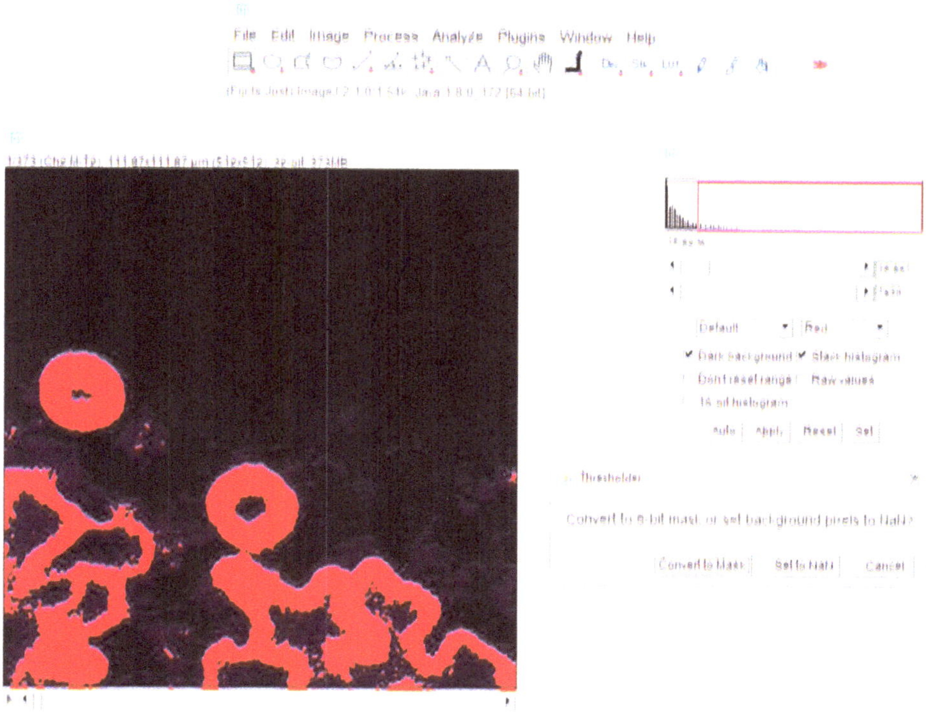

Fig. 4 Screenshot of the Fiji interface with selected intensity threshold. The "Threshold Manager window" displays the pixel intensity histogram, with the selected intensity range highlighted by a red square. After confirming the selection with "Apply," the "Threshold Conversion Options" window appears below. In the image on the left, the selected signal is displayed in red, while the background is shown in black

in black) (Fig. 4). Click "Apply" and convert the background pixels to NaN (not a number).

8. The image appears as a red and black representation. To make intensity changes detectable to the human eye, select an appropriate lookup table. Navigate to "Image" > "Lookup Tables" and choose an option, e.g., "Red Hot".
9. Define every guard cell as an ROI. Open the ROI manager, go to "Analyze" > "Tools" > "Roi Manager". To mark a specific guard cell, use the polygon selection tool in the toolbar. After drawing the region, add it to the ROI Manager by clicking on "Add". Multiple ROIs can be drawn at positions of choice. The ROIs are numbered based on their selection order (*see* **Note 19**).
10. Measure the mean gray value (fluorescence intensity) for the selected ROIs. First, define the parameter which will be measured by going to "Analyze" > "Set Measurements..." > and select "Mean gray value" > "OK."

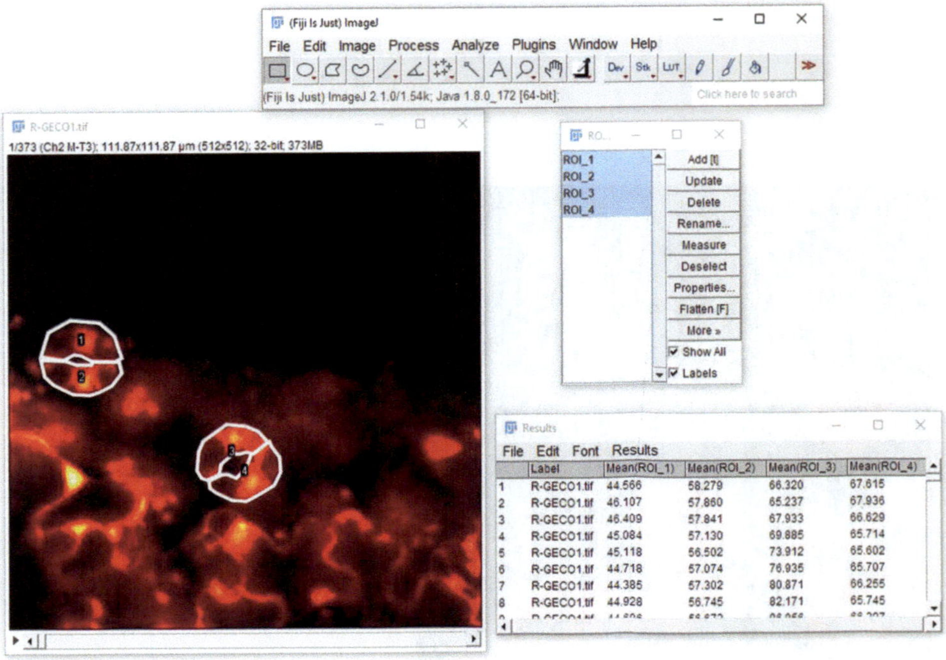

Fig. 5 Screenshot of the Fiji interface with selected ROIs. The "ROI Manager" window stores the saved ROIs for selected guard cells. The ROIs are displayed on the image at the left. The "Results" window shows the Mean gray values for the selected ROIs for every frame

11. To measure the mean gray value in the different guard cells, click in the ROI Manager on "More" > select "Multi Measure". In the Multi Measure window, select "Measure all slices," and "One row per slice..." > "OK". The Results window shows the mean fluorescence intensity for the selected ROIs over time (Fig. 5). Results are further processed in Excel.
12. To calculate FRET intensity ratio, divide the mean gray values from cpVenusFRET (CFP excitation/YFP emission) by the mean gray values of the CFP (CFP excitation/CFP emission) image series.
13. To normalize the FRET intensity ratios (R), divide R for every time point by the mean FRET intensity ratios of the 10 frames before treatment (R0). Normalize the R-GECO1 fluorescence gray value (F) equivalently by dividing F for every time point by the mean fluorescence gray of the 10 frames before treatment (F0).
14. For intensity-over-time plots the FRET intensity ratio (R/R0) or the R-GECO1 fluorescence (F/F0) is used for the y-axis values (*see* **Note 20**). The time based on the time interval between subsequent frames is used as x-axis values.

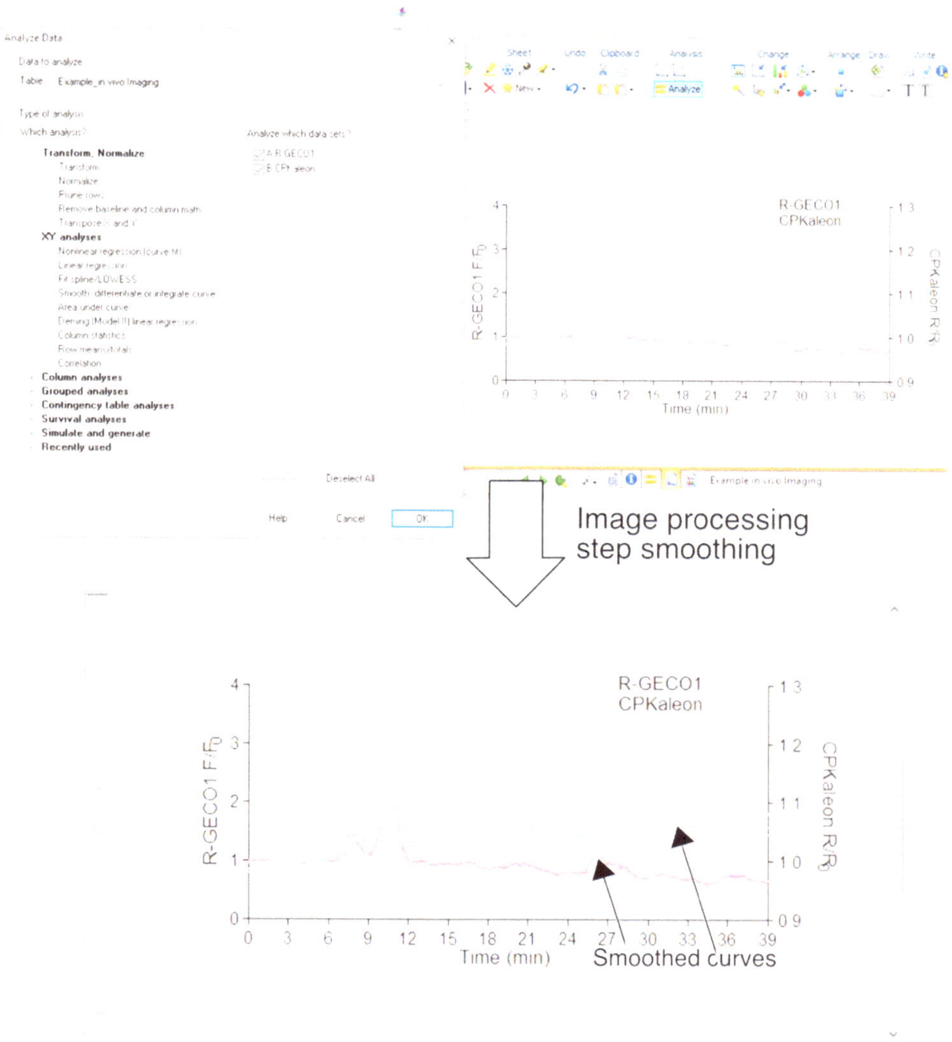

Fig. 6 Screenshot of the GraphPad interface showing the data processing step "smoothing". Data are represented in an XY chart right window. The "Analyze Data" window shows data processing options. The curve based on smoothed data is shown below

15. X- and Y-values are further processed in GraphPad Prism 5. Enter X- and Y values in one X-Y chart (Fig. 6).

16. In GraphPad Prism 5, navigate to "Analyze" > "XY analyses" > "Smooth, Differentiate, or Integrate Curve". In the "Parameters: Smooth, Differentiate, or Integrate Curve" window, set the following options: (i) Under "Differentiating or Integrating," select "Don't differentiate or integrate". (ii) Under "Smooth," choose "15 neighbors on each side" and "2nd order of the smoothing polynomial" (*see* **Note 21**).

    Click "OK" and verify that the smoothed curve is plotted on the right y-axis.

Synchronization analysis using time-dependent changes in FRET intensity ratio (R/R0) or R-GECO1 fluorescence (F/F0) is performed in R using RStudio as graphical user interface (GUI). This analysis allows an insight into the temporal resolution and periodicity of the two signals (excluding the amplitude). Before synchronization analysis the data can be revised from signal changes derived from technical artefacts such as those occurring when stomata are closing during the measurement by calculating a trend line.

17. Open RStudio program and load the R script "analysis_R-GECO1_CDPK-FRET.R" by "File" > "Open File" .... The R script and sample data are available on a github repository (https://github.com/ipb-halle/analysis_R-GECO1_CDPK-FRET).
18. Setting of the working directory which includes the Excel file "GC.xlsx" with separate columns (column name) for time (Time), recorded frame number (Frames), normalized R-GECO1 fluorescence (RG) and normalized FRET intensity ratio (FRET_ratio) of the measurement. The script loads and installs (if necessary) the required packages "readxl" [30], "writexl" [31], "scales" [32], "stats" [33], and "forecast" [34], loads data from Excel file "GC.xlsx" and creates a results folder in the selected directory (*see* **Note 22**). After adjusting the parameters start the R script by run line by line.

```
rm(list = ls())

setwd("path to the working directory")
requirements=c("readxl","writexl","scales","stats","fore-
cast")
missing <- setdiff(requirements,
 rownames(installed.packages()))

if (length(missing) != 0)
 install.packages(missing, dependencies = TRUE)
}
invisible(lapply(requirements, require, character.only =
TRUE))
dir.create(paste(getwd(),"/results",sep=""))
table <- read_excel("GC.xlsx")
```

19. In this step a linear regression model can be applied with time as the predictor variable and signal (normalized FRET intensity ratio and normalized R-GECO1 fluorescence) as the response variable to create a trend line, assuming that the technical artefacts can be represented by a linear relationship. The values of the linear regression model were used to normalize the data

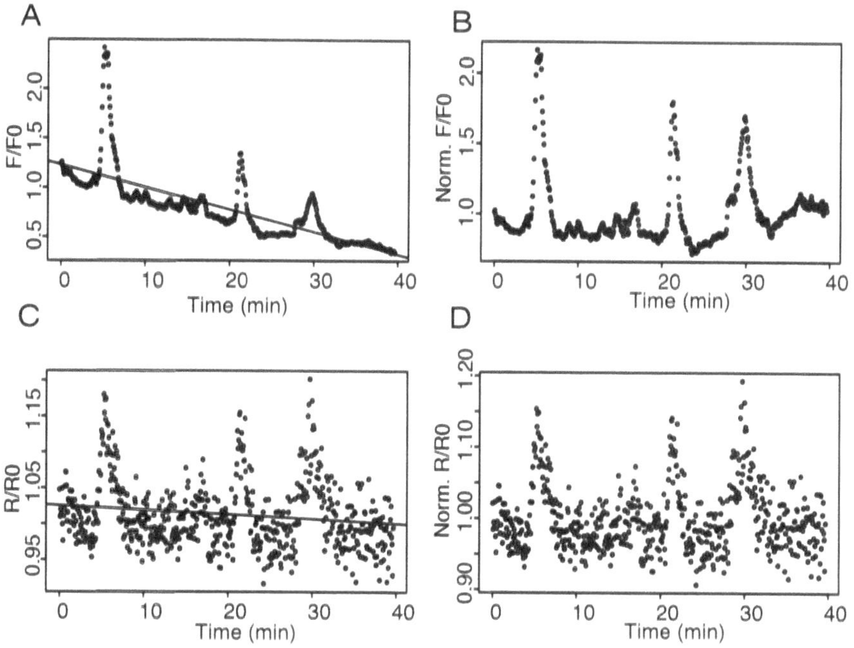

Fig. 7 Correction of R-GECO1 fluorescence and FRET intensity ratio using a linear regression model for normalization. R-GECO1 fluorescence (F/F0) and FRET intensity ratios (R/R0) before normalization with a linear regression model (line in **a** and **c**) and after normalization (**b** and **d**)

by dividing the signals by the corresponding data of the linear regression (Fig. 7, *see* **Note 23**).

```
model_RG <- lm(RG ~ Time, data = table)
plot(table$RG ~ table$Time, xlab = "Time (min)", ylab = "F/
F0")
abline(model_RG)
model_FRET <- lm(FRET_ratio ~ Time, data = table)
plot(table$FRET_ratio ~ table$Time, xlab = "Time (min)", ylab
= "R/R0")
abline(model_FRET)
lm_RG <- (table$Time * model_RG$coefficients[-c(1)]) + mod-
el_RG$coefficients[c(1)]
lm_FRET <- (table$Time * model_FRET$coefficients[-c(1)]) +
 model_FRET$coefficients[c(1)]

table$normalized_RG <- table$RG/lm_RG
table$normalized_FRET <- table$FRET_ratio/lm_FRET

plot(table$normalized_RG ~ table$Time, xlab = "Time (min)",
ylab = "Norm. F/F0")
plot(table$normalized_FRET ~ table$Time, xlab = "Time (min)",
ylab = "Norm. R/R0")
```

20. Normalized data for R-GECO1 fluorescence and normalized FRET intensity ratio were smoothed by using a moving average function. This function calculates the average of 8 neighboring data points (4 before and 4 after the actual data point). This method can be used for time series data enhancing the signal-to-noise ratio to emphasize the signal change (*see* **Note 21**).

```
table$RG_rescaled <- NA
table$FRET_rescaled <- NA
smoothed_normalized_RG <- as.numeric(ma(table$normalized_RG,
order = 8,
 centre = TRUE))
smoothed_normalized_FRET <- as.numeric(ma(table$normalized_-
FRET, order = 8,
 centre = TRUE))
```

21. For the direct comparison of the area under the curve between R-GECO1 and CPKaleon the smoothed data are re-scaled between 0 and 1 after treatment application (frame 45) by setting the minimum value to 0 and the maximum to 1 (Fig. 8).

```
table$RG_rescaled[45:(nrow(table) - 4)] <- rescale(smoothed_-
normalized_RG
 [45:(nrow(table) - 4)], c(0,1))
table$FRET_rescaled[45:(nrow(table) - 4)] <- rescale(smooth-
ed_normalized_FRET
 [45:(nrow(table) - 4)], c(0,1))
plot(table$RG_rescaled[45:(nrow(table) - 4)] ~ table$Time[45:
(nrow(table) - 4)],
 xlab = "Time (min)", ylab = "Rescaled norm. F/F0")
plot(table$FRET_rescaled[45:(nrow(table) - 4)] ~ table$Time
[45:(nrow(table) - 4)],
 xlab = "Time (min)", ylab = "Rescaled norm. R/R0")
```

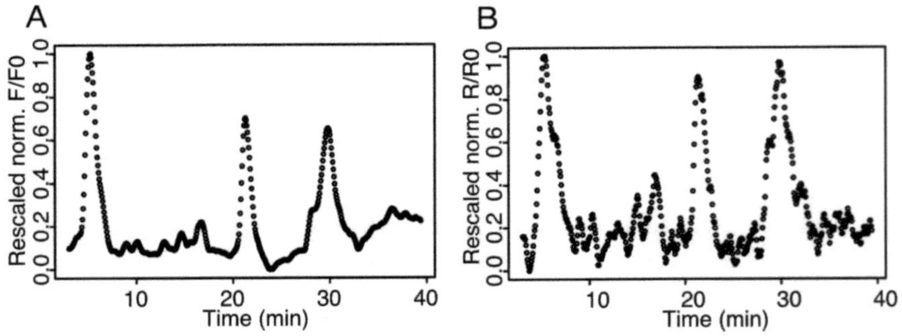

Fig. 8 Rescaling of R-GECO1 fluorescence and FRET intensity ratios. The plots showing the normalized, smoothed and rescaled data after treatment application for the R-GECO1 fluorescence (**a**) and FRET intensity ratios (**b**)

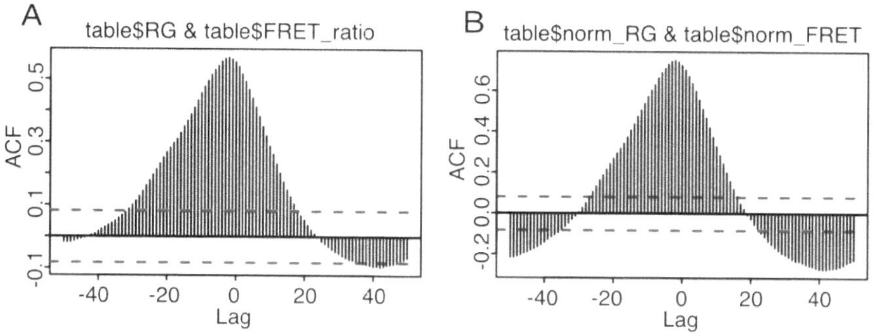

Fig. 9 Synchronization analysis of R-GECO1 fluorescence and FRET intensity ratio. Analyses show the correlation coefficient (ACF) dependent on the lag phase between the R-GECO1 fluorescence and FRET intensity ratio before (**a**) and after normalization (**b**)

22. The synchronization analysis is performed by calculating a cross correlation (Fig. 9) between the R-GECO1 fluorescence and FRET intensity ratio before (ccf) and after linear regression normalization (ccf_normalized) for a maximal lag phase of 50 frames (in this measurement this corresponds to approximately 3.5 min, *see* **Notes 24**).

```
max_lag = 50
cross_corr_matrix = matrix(0,nrow=105,ncol = 4)
colnames(cross_corr_matrix) <- c("lag", "Time", "Correlation
coefficients",
 "Correlation coefficients normalized")

ccf <- ccf(table$RG, table$FRET_ratio, lag.max = max_lag)
ccf_normalized <- ccf(table$normalized_RG, table$normalized_-
FRET, lag.max = max_lag)
```

23. Create and export a cross-correlation matrix including the correlation coefficients of the cross-correlation before and after normalization for a lag phase of −50 to +50 frames and summarizes the maximal correlation coefficient and the corresponding lag phase in minutes ("cross correlation.xlsx").

```
cross_corr_matrix[1:101,1] = ccf$lag
cross_corr_matrix[1:101,2] = ccf$lag * table$Time[2]
cross_corr_matrix[1:101,3] = ccf$acf
cross_corr_matrix[1:101,4] = ccf_normalized$acf
cross_corr_matrix[102,1] = "lag max Time ccf"
cross_corr_matrix[103,1] = "max correlation coefficient cff"
cross_corr_matrix[102,2] = ccf$lag[which.max(cross_corr_ma-
trix[1:101,3])] *
 table$Time[2]
cross_corr_matrix[103,2] = ccf$acf[which.max(cross_corr_ma-
```

```
trix[1:101,3])]
cross_corr_matrix[104,1] = "lag max Time ccf normalized"
cross_corr_matrix[105,1] = "max correlation coefficient cff
normalized"
cross_corr_matrix[104,2] = ccf_normalized$lag[which.max
(cross_corr_matrix[1:101,4])] *
 table$Time[2]
cross_corr_matrix[105,2] = ccf_normalized$acf[which.max
(cross_corr_matrix[1:101,4])]
write_xlsx(data.frame(cross_corr_matrix),
 path=paste(getwd(),"/results/cross correlation.xlsx",
sep=""), col_names = TRUE)
```

24. Export of the original data frame including additional columns for the normalized data ("normalized_RG", "normalized_-FRET") and the rescaled, smoothed data ("RG_rescaled", "FRET_rescaled") as Excel sheet ("data.xlsx").

```
write_xlsx(data.frame(table),path=paste(getwd(),"/results/
data.xlsx",sep=""), col_names = TRUE)
```

25. To calculate the "area under the curve" the data is further processed in GarphPad Prism. The normalized, smoothed and rescaled data after treatment application for the R-GECO1 fluorescence and FRET intensity ratios are displayed as X-Y chart.
26. In GraphPad Prism 5, navigate to "Analyze" > "XY analyses" > "Area under the curve." In the "Parameters: Area under the Curve" window, set the following options: "Baseline Y = 0.0" under "Minimum peak height," select "Ignore peaks that are Less than 10 % of distance from minimum to maximum Y" under "Peak direction" choose "By definition, all peaks must go above the baseline" and under "Significant digits" select "Show: 4."
27. Click "OK."

4 Notes

1. For the buffer with a high imidazole concentration, it is important to adjust the pH after adding imidazole, as the acidic proton of imidazole causes pH changes.
2. The four acidic groups of EGTA are in equilibrium with both hydrogen ions and metal ions. The various deprotonated forms of EGTA have different dissociation constants for calcium. Therefore, it is crucial to adjust the pH properly to ensure accurate calcium buffering.

3. Even though we use a high concentration of EGTA to bind contaminating calcium, it is impossible to reduce the calcium concentration to zero. Since the exact contaminating calcium concentration is unknown and cannot be calculated, we refer to the buffer as a "zero-Ca^{2+} buffer."
4. When Ca^{2+} binds to EGTA, calcium is exchanged for H^+, resulting in an increase in free H^+ concentration and a decrease in pH. Therefore, it is essential to readjust the pH.
5. ABA has a higher solubility in an organic solvent than in water. Therefore, 10 mM ABA is dissolved in 100 % ethanol and diluted to 1 mM with water.
6. The doubling time of the cell culture should be approximately 30 min.
7. The CPK21-based CPKaleon variant has a molecular weight (MW) of approximately 120 kDa, equivalent to 120,000 g/mol. An optimal protein concentration in FRET measurements prior to a 1:1 dilution is 415 nM, or 0.0498 g/L.
8. For the calculation of the free calcium concentration, it should be noted that for the measuring mix the calcium-EGTA buffer was 1:1 diluted with CPKs in dialysis buffer.
9. At the pH 7.4 used for the in vitro FRET measurement, 30 mM Tris (pKa = 8.2) dissociates in 25.9 mM Tris and 4.1 mM $TrisH^+$.
10. At the pH of 7.4 used for the in vitro FRET measurements, EGTA ($pKa_1 = 9.43, pKa_2 = 8.85, pKa_3 = 2.68, pKa_4 = 2.0$) is essentially completely in double negative-charged form.
11. We use GraphPad Prism 5 software for curve fitting of in vitro FRET measurements. This is a very user-friendly program for this purpose, but also other suitable programs exist.
12. In in vitro FRET measurements, our goal is to find the EC_{50} value. In contrast, the HillSlope is less informative. Many programs enable to create a global model for the HillSlope combining several data sets for the same CPKaleon isoform. This approach enhances the certainty of the individual fitted EC_{50} values.
13. Uneven expression observed in β-estradiol inducible system restricts the number of reporter expressing guard cells per imaging area.
14. By opening the pinhole, the signal increases; however, signals from the out-of-focus planes also increase. This negative side effect can be neglected in a single-cell layer system.
15. If a high pixel frequency is detected in the high-intensity range of the detector before treatment, even a small increase in pixel intensity can lead to detector saturation. Therefore, fluorescence increases cannot be detected accurately.

16. A 16-bit grayscale format captures 65,536 shades of gray, enabling the depiction of even minor changes in gray intensity. To reduce file size, we use a frame size of 512 × 512 pixels as a compromise between the accuracy of digitally encoded information and manageable data handling.
17. Application of volumes larger than 50 μL increases the risk of accidental sample movement. Therefore, it is helpful to use a 50-fold concentrated stock for small-volume stimulus application.
18. For ABA in 10% ethanol, perform a control measurement with 10% ethanol.
19. We recommend saving the ROIs as selections to document analysis progress. To save a selected ROI, click on it in your ROI Manager and navigate to "File" > "Save As" > "Selection." For the different fluorescence channels of the same time series, use the same selection for analysis.
20. The maximum R-GECO1 change is higher than the maximum FRET Ratio change. Therefore, we plot both of them on two y-axis (left/right) to visualize them in only one diagram.
21. Technical artefacts and local calcium changes may lead to imaging noise. Therefore, smoothing imaging data sets may help to capture important patterns.
22. The Excel file can be renamed, but this name must be adjusted in line 21 of the R script. We recommend creating a folder for each measurement and setting the working directory to this folder to prevent overwriting of previous results.
23. In the case that the measurement does not show artefacts like in general decreasing fluorescence (Fig. 7) due to movement of the sample, this step is not recommended (the synchronization analysis is performed with and without normalization by the linear regression model).
24. Comparing longer lag phases may be useful to discover repetitive elements in the synchronization analysis like oscillating signatures. When the lag time (line 55) is changed in the R Script, the matrix (line 56) of the ccf-output table also needs to be adjusted (lines 61–72).

Acknowledgments

This work was supported by the Deutsche Forschungsgemeinschaft (DFG—514901783) through the Collaborative Research Centre 1664 "Plant Proteoform Diversity-SNP2Prot" (projects B02/B03).

References

1. Asai T, Tena G, Plotnikova J, Willmann MR, Chiu WL, Gomez-Gomez L, Boller T, Ausubel FM, Sheen J (2002) MAP kinase signalling cascade in *Arabidopsis* innate immunity. Nature 415:977–983
2. Seybold H, Boudsocq M, Romeis T (2017) CDPK activation in PRR signaling. Methods Mol Biol 1578:173–183
3. Suzuki K, Shinshi H (1995) Transient activation and tyrosine phosphorylation of a protein kinase in tobacco cells treated with a fungal elicitor. Plant Cell 7:639–647
4. Wang P, Zhu JK (2016) Assessing kinase activity in plants with in-gel kinase assays. Methods Mol Biol 1363:189–197
5. Brumbaugh J, Schleifenbaum A, Gasch A, Sattler M, Schultz C (2006) A dual parameter FRET probe for measuring PKC and PKA activity in living cells. J Am Chem Soc 128: 24–25
6. Depry C, Zhang J (2011) Using FRET-based reporters to visualize subcellular dynamics of protein kinase A activity. Methods Mol Biol 756:285–294
7. Zaman N, Seitz K, Kabir M, George-Schreder LS, Shepstone I, Liu Y, Zhang S, Krysan PJ (2019) A Förster resonance energy transfer sensor for live-cell imaging of mitogen-activated protein kinase activity in Arabidopsis. Plant J 97:970–983
8. Zhang L, Takahashi Y, Hsu PK, Kollist H, Merilo E, Krysan PJ, Schroeder JI (2020) FRET kinase sensor development reveals SnRK2/OST1 activation by ABA but not by MeJA and high CO_2 during stomatal closure. Elife 9:e56351
9. Cheng SH, Willmann MR, Chen HC, Sheen J (2002) Calcium signaling through protein kinases. The Arabidopsis calcium-dependent protein kinase gene family. Plant Physiol 129: 469–485
10. Miyawaki A, Llopis J, Heim R, McCaffery JM, Adams JA, Ikura M, Tsien RY (1997) Fluorescent indicators for Ca^{2+} based on green fluorescent proteins and calmodulin. Nature 388: 882–887
11. Liese A, Eichstädt B, Lederer S, Schulz P, Oehlschläger J, Matschi S, Feijó JA, Schulze WX, Konrad K, Romeis T (2023) Imaging of plant calcium-sensor kinase conformation monitors real time calcium-dependent decoding *in planta*. Plant Cell 36:276–297
12. Crooks GE, Hon G, Chandonia JM, Brenner SE (2004) WebLogo: a sequence logo generator. Genome Res 14:1188–1190
13. Goedhart J, van Weeren L, Hink MA, Vischer NOE, Jalink K, Gadella TWJ Jr (2010) Bright cyan fluorescent protein variants identified by fluorescence lifetime screening. Nat Methods 7:137–139
14. Heim N, Griesbeck O (2004) Genetically encoded indicators of cellular calcium dynamics based on troponin C and green fluorescent protein. J Biol Chem 279:14280–14286
15. Nagai T, Yamada S, Tominaga T, Ichikawa M, Miyawaki A (2004) Expanded dynamic range of fluorescent indicators for Ca^{2+} by circularly permuted yellow fluorescent proteins. Proc Natl Acad Sci U S A 101:10554–10559
16. Klimecka M, Muszyńska G (2007) Structure and functions of plant calcium-dependent protein kinases. Acta Biochim Pol 54:219–233
17. Gifford JL, Walsh MP, Vogel HJ (2007) Structures and metal-ion-binding properties of the Ca^{2+}-binding helix-loop-helix EF-hand motifs. Biochem J 405:199–221
18. Sigrist CJ, de Castro E, Cerutti L, Cuche BA, Hulo N, Bridge A, Bougueleret L, Xenarios I (2013) New and continuing developments at PROSITE. Nucleic Acids Res 41:D344–D347
19. Weber E, Engler C, Gruetzner R, Werner S, Marillonnet S (2011) A modular cloning system for standardized assembly of multigene constructs. PLoS One 6:e16765
20. McClendon CL, Kornev AP, Glison MK, Taylor SS (2014) Dynamic architecture of a protein kinase. Proc Natl Acad Sci U S A 111: E4623–E4631
21. Studier FW, Moffatt BA (1986) Use of bacteriophage T7 RNA polymerase to direct selective high-level expression of cloned genes. J Mol Biol 189:113–130
22. Bers DM, Patton CW, Nuccitelli R (2010) A practical guide to the preparation of Ca^{2+} buffers. Methods Cell Biol 99:1–26
23. Waadt R, Krebs M, Kudla J, Schumacher K (2017) Multiparameter imaging of calcium and abscisic acid and high-resolution quantitative calcium measurements using R-GECO1-mTurquoise in Arabidopsis. New Phytol 216: 303–320
24. Green R, Rogers EJ (2013) Transformation of chemically competent *E. coli*. In: Lorsch J (ed) Methods in enzymology, vol 529. Academic, San Diego, pp 329–336
25. Kurien BT, Scofield RH (2015) Multiple immunoblots by passive diffusion of proteins from a single SDS-PAGE gel. Methods Mol Biol 1312:77–86

26. Borghi L (2010) Inducible gene expression systems for plants. Methods Mol Biol 655: 65–75
27. Schindelin J, Arganda-Carreras I, Frise E, Kaynig V, Longair M, Pietzsch T, Preibisch S, Rueden C, Saalfeld S, Schmid B, Tinevez JY, White DJ, Hartenstein V, Eliceiri K, Tomancak P, Cardona A (2012) Fiji: an open-source platform for biological-image analysis. Nat Methods 9:676–682
28. R Core Team (2023) R: A language and environment for statistical computing. R Foundation for Statistical Computing, Vienna
29. Li K, Prada J, Damineli DSC, Liese A, Romeis T, Dandekar T, Feijó JA, Hedrich R, Konrad KR (2021) An optimized genetically encoded dual reporter for simultaneous ratio imaging of Ca^{2+} and H^+ reveals new insights into ion signaling in plants. New Phytol 230: 2292–2310
30. Wickham H, Bryan J RStudio Team (2022) readxl: Read Excel Files. R package version 1.4.1
31. Ooms J (2021) writexl: Export data frames to Excel '.xlsx' format. R package version 1.4.0
32. Wickham H, Seidel D, RStudio Team (2022) scales: Graphical scales for visualization. R package version 1.2.1
33. Hyndman RJ, Khandakar Y (2008) Automatic time series forecasting: the forecast package for R. J Stat Softw 27:1–22
34. Hyndman R, Athanasopoulos G, Bergmeir C, Caceres G, Chhay L, O'Hara-Wild M, Petropoulos F, Razbash S, Wang E, Yasmeen F, RStudio (2022) forecast: Forecasting functions for time series and linear models. R package version 8.17

Chapter 6

Visualization and Quantification of Apoplastic ROS by Plant Immune NADPH Oxidase Utilizing Chemiluminescence

Miki Yoshioka and Hirofumi Yoshioka

Abstract

Reactive oxygen species (ROS) are implicated in plant innate immunity. When plants recognize pathogens, apoplastic superoxide (O_2^-) is produced by a plasma membrane NADPH oxidase (RBOH; Respiratory Burst Oxidase Homolog), which is immediately dismutated to hydrogen peroxide (H_2O_2). *Nicotiana benthamiana* NbRBOHB is responsible for both the transient pattern-triggered immunity (PTI)-ROS burst and the robust effector-triggered immunity (ETI)-ROS burst. Potato (*Solanum tuberosum*) StRBOHB and StRBOHC are phosphorylated by potato calcium-dependent protein kinases, StCDPK4 and StCDPK5, resulting in the ROS burst. The ROS contribute to the potato resistance to a virulent isolate of late blight pathogen, *Phytophthora infestans*. Here, we introduce an apoplastic ROS detection protocol utilizing L-012-derived chemiluminescence in response to *Agrobacterium*-mediated co-expression of StRBOHC and StCDPK5 in *N. benthamiana* leaves.

Key words Apoplastic ROS, NADPH oxidase, Calcium-dependent protein kinase, Chemiluminescence, L-012, RBOH

1 Introduction

Rapid production of reactive oxygen species (ROS) has been implicated in plant innate immunity [1]. The NADPH oxidases, RBOHs (respiratory burst oxidase homologs), play central roles in ROS burst in the plant apoplast. Plants exhibit a two-phase immune system to protect themselves from threats such as pathogenic fungi and bacteria. Plants recognize pathogen-associated molecular patterns (PAMPs) by pattern recognition receptors (PRRs) at the plasma membrane, and initiate a pattern-triggered immunity (PTI), mostly this recognition occurs at apoplastic regions. The other is effector-triggered immunity (ETI), a robust reaction that recognizes pathogen effectors and is accompanied by cell death. In *Nicotiana benthamiana*, NbRBOHB is responsible for both the transient PTI-ROS burst and the robust ETI-ROS burst [2].

Tsutomu Kawasaki and Hirofumi Yoshioka (eds.), *Plant Immunity: Methods and Protocols*, Methods in Molecular Biology, vol. 3012, https://doi.org/10.1007/978-1-0716-5138-4_6,

RBOH-mediated ROS in the apoplastic region influx into the cytosol via H_2O_2-permeable aquaporins [3], and sulfenylate the cysteine residues of ROS sensor proteins responsible for the plant immune responses during PTI and ETI [4]. Phosphorylation of N-terminal region of *Arabidopsis* AtRBOHD by receptor-like cytoplasmic kinase (RLCK) during PTI induces PTI-ROS burst and immune responses [5]. Potato (*Solanum tuberosum*) calcium-dependent protein kinase 5 (StCDPK5) has been also shown to phosphorylate the N-terminal region of StRBOHB, and provokes the ROS burst [6]. CDPK is a serine/threonine protein kinase and consists of four domains: an N-terminal variable domain (V), a protein kinase domain (K), an autoinhibitory junction domain, and a C-terminal calmodulin-like domain (CaM-LD) including four EF-hand motifs, which are pre-bounding to the junction at basal levels of Ca^{2+} [7]. A truncated VK form, removing a junction and CaM-LD from CDPK, shows constitutive activity in the absence of Ca^{2+} [8]. Transient heterologous co-expression of StCDPK5VK and StRBOHA, StRBOHB, StRBOHC, or StRBOHD in *N. benthamiana* induced a robust apoplastic ROS burst [6, 9] and the stable transgenic potato plants, harboring StCDPK5VK under control of the pathogen-inducible promoter, showed high resistance to a virulent isolate of *P. infestans* infection [10].

Agrobacterium-mediated transient gene expression in *N. benthamiana* is a simple and reproducible experimental system. The mixture of transformed *Agrobacterium* suspensions with StRBOHB and StCDPK5VK or kinase-inactive StCDPK5KM was infiltrated into the leaves (Fig. 1). The generated ROS were detected as a chemiluminescence using L-012 probe [11]. The L-012 is a highly reactive reagent for O_2^- detection without peroxidase in vitro, and also reacts with H_2O_2 using peroxidase [12]. The emitted light can be captured by a highly sensitive CCD camera, and visualized the ROS distribution in the leaves (Fig. 2). The L-012-derived chemiluminescence in response to a PAMP and an effector is abolished by NbRBOHB-silenced *N. benthamiana* leaves [2]. Therefore, the reagent is suitable for visualization and quantification of RBOH-mediated ROS in the apoplastic region of the leaves during PTI and ETI [2].

2 Materials

Unless otherwise noted, deionized water (Millipore Elix) and basic molecular biology reagents and equipment are used to prepare all of the following solutions.

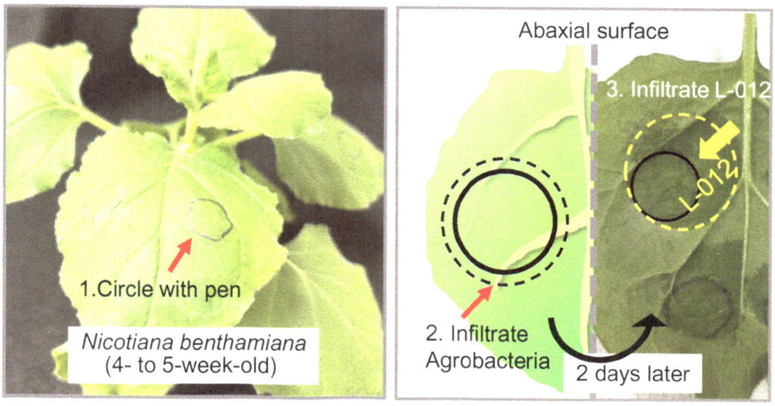

Fig. 1 Injection of L-012 into *N. benthamiana* leaves. The fully expanded fourth to sixth leaves of 4- to 5-week-old *N. benthamiana* are used for experiments. (1) Using an oil-based pen, draw two circles as close to the same size as possible on the adaxial side of the leaf, avoiding the veins. (2) Infiltrate the *Agrobacterium* suspension with a needleless syringe, slightly beyond the marked circle from the abaxial side of the leaf. (3) Two days after *Agrobacterium* injection, infiltrate L-012 from the abaxial side of the leaf with a needleless syringe, from the outside of the circle to the center of the circle

2.1 Agrobacterium-Mediated Transient Expression

1. 4- to 5-week-old *N. benthamiana*.
2. *Agrobacterium tumefaciens* strain GV3101 containing 35S: flag-StRBOHC in pGD binary plasmid [13] and constitutively active 35S:StCDPK5VK-HA-StrepII or kinase-inactive 35S: StCDPK5KM-HA-StrepII mutants in pGreen binary plasmid [14].
3. LB media.
4. Suspension buffer: 10 mM MES/NaOH, 10 mM $MgCl_2$, pH 5.6, sterilize by filtration.
5. 150 mM acetosyringone dissolved in DMSO, aliquot, and store at −20 °C.
6. 1 mL syringe.

2.2 ROS Detection and Data Analysis

1. 20 mM L-012 (FUJIFILM Wako Chemicals) in deionized water, 30 μL aliquot in 1.5 mL tube, and store at −20 °C.
2. 500 mM MOPS-KOH, pH 7.4, sterilize by filtration.
3. LuminoGraph III (ATTO).
4. CS Analyzer 4 software to analyze chemiluminescence signals (ATTO).

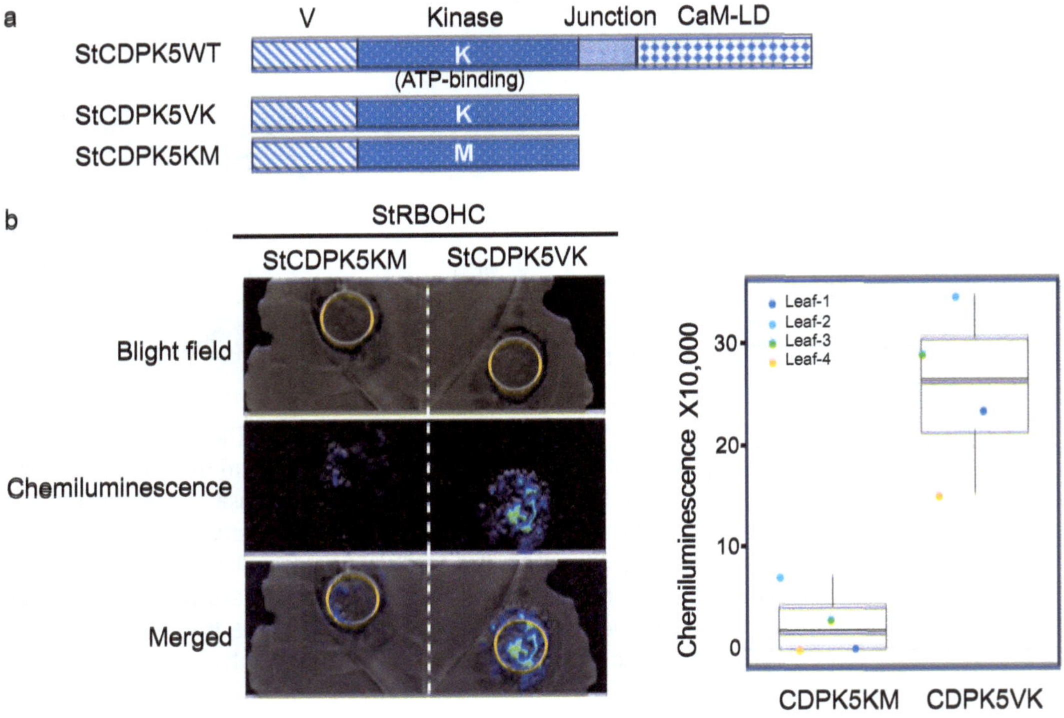

Fig. 2 Visualization and quantification of ROS induced by co-expression of RBOH and CDPK. Schematic structures of StCDPK5 variants (**a**). "V" indicates variable domain, "Kinase" indicates kinase domain, "Junction" indicates junction domain, "CaM-LD" indicates calmodulin-like domain, and "K" indicates a Lys residue for ATP binding in the kinase domain. VK is a truncated variant lacking Junction and CaM-LD. KM is a truncated kinase-inactive variant with substitution of Lys (K) in the ATP-binding motif with Met (M). (**b**) ROS production was detected as chemiluminescence using L-012 probe. L-012-infiltrated leaf was placed under the CCD camera with adaxial surface up. The emitted chemiluminescence was detected by LuminoGraph III (ATTO) with BPF595 filter, 6 × 6 binning for 5 min, and analyzed using a CS Analyzer 4 software (ATTO)

3 Methods

3.1 Agrobacterium-Mediated Transient Expression

1. Inoculate one single colony of *Agrobacterium* containing the 35S:flag-StRBOHC, 35S:StCDPK5VK, and kinase-inactive mutant, 35S:StCDPK5KM constructs in 2 mL LB with appropriate antibiotics overnight at 28 °C, shaking at 140 rpm.
2. Use 2 mL of the overnight culture to inoculate 20 mL LB with the same antibiotics and incubate for 5 h, at 28 °C, shaking at 140 rpm.
3. Precipitate the *Agrobacterium* (3000× *g*, 15 min), then resuspend the pellet in 10 mL of suspension buffer.
4. Precipitate the *Agrobacterium* (3000× *g*, 15 min), then resuspend the pellet in 5 mL of suspension buffer containing 150 μM acetosyringone.
5. Measure the optical density (OD) at 600 nm, and adjust to 0.5.

6. Incubate the bacterial suspension at 23 °C for 2 h, and suspensions containing StRBOHC and StCDPK5VK or StCDPK5KM were mixed in a ratio of 4:1.
7. Mark two circles with a marker pen on the adaxial surface of *N. benthamiana* leaf.
8. Infiltrate each suspension into one circle from the abaxial side of the leaf using a needleless syringe (*see* **Note 1**).

3.2 ROS Detection

Two days after infiltration, the relative intensity of ROS generation is determined by detection of L-012–mediated chemiluminescence.

1. 24 μL of 500 mM MOPS-KOH, pH 7.4 from Subheading 2.2, **item 2** and 1146 μL deionized water are added into L-012 aliquot tube from Subheading 2.2, **item 1**, which makes a final concentration of 0.5 mM L-012 and 10 mM MOPS-KOH, pH 7.4.
2. Infiltrate the L-012 mixture into the circle from the abaxial side of the leaf, which has been co-expressed StRBOHC and StCDPK5 variants (*see* **Note 2**).
3. Detect the chemiluminescence from the adaxial surface of the leaves using LuminoGraph III (ATTO) equipped with a highly sensitive CCD camera by BPF595 filter, 6 × 6 binning for 5 min.
4. Using the CS Analyzer 4 software (ATTO), each intensity of chemiluminescence is calculated by spot area analysis.

4 Notes

1. For *Agrobacterium* infection, infiltrate the suspension cells slightly beyond the area of the marked circle from the abaxial surface.
2. L-012 is injected from slightly outside the circle of the abaxial side of the leaf toward the center of the circle to avoid wounding damage. This is because the wounding with a syringe may prevent the detection of ROS.

Acknowledgments

We thank Andrew O. Jackson for pGD vector and Phil Mullineaux and Roger Hellens for pGreen vector. We also thank the Leaf Tobacco Research Center for *N. benthamiana* seeds. This research was supported by Grants-in-Aid for Exploratory Research (25K22359) to H.Y.

References

1. Doke N (1983) Involvement of superoxide anion generation in the hypersensitive response of potato tuber tissues to infection with an incompatible race of *Phytophthora infestans* and to the hyphal wall components. Physiol Plant Pathol 23:345–357
2. Adachi H, Nakano T, Miyagawa N, Ishihama N, Yoshioka M, Katou Y, Yaeno T, Shirasu K, Yoshioka H (2015) WRKY transcription factors phosphorylated by MAPK regulate a plant immune NADPH oxidase in *Nicotiana Benthamiana*. Plant Cell 27:2645–2663
3. Tian S, Wang X, Li P, Wang H, Ji H, Xie J, Qiu Q, Shen D, Dong H (2016) Plant aquaporin AtPIP1;4 links apoplastic H_2O_2 induction to disease immunity pathways. Plant Physiol 171:1635–1650
4. Hino Y, Inada T, Yoshioka M, Yoshioka H (2024) NADPH oxidase-mediated sulfenylation of cysteine derivatives regulates plant immunity. J Exp Bot 75:4641–4654
5. Kadota Y, Sklenar J, Derbyshire P, Stransfeld L, Asai S, Ntoukakis V, Jones JDG, Shirasu K, Menke F, Jones A, Zipfel C (2014) Direct regulation of the NADPH oxidase RBOHD by the PRR-associated kinase BIK1 during plant immunity. Mol Cell 54:43–55
6. Kobayashi M, Ohura I, Kawakita K, Yokota N, Fujiwara M, Shimamoto K, Doke N, Yoshioka H (2007) Calcium-dependent protein kinases regulate the production of reactive oxygen species by potato NADPH oxidase. Plant Cell 19:1065–1080
7. Harper JF, Breton G, Harmon A (2004) Decoding $Ca^{(2+)}$ signals through plant protein kinases. Annu Rev Plant Biol 55:263–288
8. Harper JF, Huang JF, Lloyd SJ (1994) Genetic identification of an autoinhibitor in CDPK, a protein kinase with a calmodulin-like domain. Biochemistry 3323:7267–7277
9. Asai S, Ichikawa T, Nomura H, Kobayashi M, Kamiyoshihara Y, Mori H, Kadota Y, Zipfel C, Jones JDG, Yoshioka H (2013) The variable domain of a plant calcium-dependent protein kinase (CDPK) confers subcellular localization and substrate recognition for NADPH oxidase. J Biol Chem 288:14332–14340
10. Kobayashi M, Yoshioka M, Asai S, Nomura H, Kuchimura K, Mori H, Doke N, Yoshioka H (2012) StCDPK5 confers resistance to late blight pathogen but increases susceptibility to early blight pathogen in potato via reactive oxygen species burst. New Phytol 196:223–237
11. Nishinaka Y, Aramaki Y, Yoshida H, Masuya H, Sugawara T, Ichimori Y (1993) A new sensitive chemiluminescence probe, L-012, for measuring the production of superoxide anion by cells. Biochem Biophys Res Commun 193:554–559
12. Asai S, Ohta K, Yoshioka H (2008) MAPK signaling regulates nitric oxide and NADPH oxidase-dependent oxidative bursts in *Nicotiana benthamiana*. Plant Cell 20:1390–1406
13. Goodin MM, Dietzgen RG, Schichnes D, Ruzin S, Jackson AO (2002) pGD vectors: versatile tools for the expression of green and red fluorescent protein fusions in agroinfiltrated plant leaves. Plant J 31:375–383
14. Hellens RP, Edwards EA, Leyland NR, Bean S, Mullineaux PM (2000) pGreen: a versatile and flexible binary Ti vector for *Agrobacterium*-mediated plant transformation. Plant Mol Biol 42:819–832

Chapter 7

Detection of Phosphorylated NADPH Oxidase by Antibody During Plant Immune Responses

Yuta Hino and Hirofumi Yoshioka

Abstract

Reactive oxygen species (ROS) production during immune responses is tightly regulated by the multiple phosphorylation of plasma membrane-localized NADPH oxidases, respiratory burst oxidase homologs (RBOHs). It is known that calcium-dependent protein kinases (CDPKs) and other kinases phosphorylate N-terminal region of RBOH and activate ROS production during immune responses. In this chapter, we describe protocols for the preparation of anti-phosphopeptide antibodies, and the detection of phosphorylated RBOH during immune responses using anti-phospho-RBOH antibody in *Nicotiana benthamiana*.

Key words Phosphorylation, Anti-phosphopeptide antibody, NADPH oxidase, ROS production

1 Introduction

Reactive oxygen species (ROS) production is a common immune response among plant species and serves as a benchmark for assessing the amplitude of immune activation. In the apoplastic region, ROS production is carried out by plasma membrane-localized NADPH oxidases, known as respiratory burst oxidase homologs (RBOHs), during both pattern-triggered immunity (PTI) and effector-triggered immunity (ETI) [1]. RBOH-dependent ROS production during immune responses is required for defense against pathogens, maintaining microbiota, and signal transduction via oxidative modification of cysteine residues [2–4]. On the other hand, excess ROS cause oxidative damage to plant cells, which means that ROS act as a double-edged sword [5]. Therefore, the activity of RBOHs must be tightly regulated to achieve spatiotemporally compartmentalized ROS accumulation.

The regulatory mechanisms of RBOH activity have been studied over the past quarter of a century. Plant RBOHs are six-transmembrane domain proteins with cytosol-exposed N- and C-terminal regions containing a Ca^{2+}-binding EF-hand motif, and

Tsutomu Kawasaki and Hirofumi Yoshioka (eds.), *Plant Immunity: Methods and Protocols*, Methods in Molecular Biology, vol. 3012, https://doi.org/10.1007/978-1-0716-5138-4_7,

an NADPH catalytic domain [6]. The activity of RBOH is modulated by multiple post-translational modifications, most notably phosphorylation [6]. Multiple kinases orchestrate fine-tuning of RBOH activity by the phosphorylation of several conserved phosphorylated residues in the N-terminal region of RBOH. For instance, receptor-like cytoplasmic kinases (RLCKs) phosphorylate several serine residues in the N-terminal cytosolic region of RBOH to trigger transient ROS production during PTI [7]. Calcium-dependent protein kinases (CDPKs) also contribute to RBOH activation in a Ca^{2+}-dependent manner. We demonstrated that StCDPK4 and StCDPK5 directly phosphorylate StRBOHB at Ser-82 and Ser-97 in the N-terminal region, thereby promoting ROS production in potato [8, 9]. In *Nicotiana benthamiana*, these serine residues are conserved in NbRBOHB Ser-123 and Ser-138, respectively [8]. The phosphorylation not only positively regulates RBOH activity, but also negatively regulates it by the C-terminal phosphorylation followed by ubiquitination and degradation of RBOH protein [10]. Thus, phosphorylation analysis of RBOHs is essential for understanding the regulatory mechanisms underlying ROS generation in plant immunity.

To facilitate this analysis, we developed a method to detect phosphorylated RBOH during immune responses using anti-phosphopeptide antibodies combined with a transient expression system via *Agrobacterium* in *N. benthamiana*. *N. benthamiana* is a powerful model system for dissecting RBOH regulation, owing to its amenability to transient expression and gene knockdown via virus-induced gene silencing (VIGS). This method is applicable not only to the study of RBOH phosphorylation during immune responses but also to other aspects of plant signaling. For example, we successfully analyzed the phosphorylation-dependent activation of rice OsRBOHH by OsCDPK5 and OsCDPK13 under hypoxic conditions [11]. While the detection of phosphorylated peptides by liquid chromatography-tandem mass spectrometry (LC-MS/MS) has become standard, it has limitations. For example, some phosphopeptides may not be detectable due to incomplete enzymatic digestion. Noteworthy, the phosphorylation of Arabidopsis AtRBOHD Ser-133 and Ser-148, which correspond to StRBOHB Ser-82 and Ser-97 phosphorylated by StCDPK4 and StCDPK5, were undetectable by LC-MS/MS due to poor yield of peptides containing these phosphorylation sites after tryptic digestion [7]. In this case, immunoblotting using phospho-specific antibodies is particularly useful for comparing phosphorylation levels across multiple samples, such as time course comparison. In this chapter, we describe a method to detect the phosphorylation of plasma membrane-localized RBOH using anti-phosphopeptide antibodies in *N. benthamiana*.

2 Materials

Unless otherwise noted, deionized water (Merck Elix) and basic molecular biology reagents and equipment are used to prepare all of the following solutions. Unless other temperatures are specified, chilled solutions can be kept on ice on the bench.

2.1 Affinity Purification of Anti-Phosphopeptide Antibodies

1. Phosphorylated and non-phosphorylated synthetic peptides (*see* **Note 1**).
2. Hitrap NHS activated HP, 1 mL (GE Healthcare).
3. Coupling buffer: 0.2 M $NaHCO_3$, pH 8.3, 0.5 M NaCl, filter through 0.45 μm filter.
4. 1 mM HCl.
5. Buffer A: 0.5 M ethanolamine-HCl, pH 8.3, 0.5 M NaCl, filter through 0.45 μm filter.
6. Buffer B: 0.1 M sodium acetate, pH 4.0, 0.5 M NaCl, filter through 0.45 μm filter.
7. Buffer C: 20 mM Tris–HCl, pH 7.4, 0.5 M NaCl, filter through 0.45 μm filter.
8. Elution buffer: 0.2 M glycine-HCl, pH 2.5, filter through 0.45 μm filter.
9. Antisera of rabbits immunized with phosphorylated peptide (*see* **Note 2**).
10. Millex-HV Syringe Filter Unit, 0.45 μm, PVDF (Millipore).
11. TBS buffer: 50 mM Tris–HCl, pH 7.5, 150 mM NaCl.
12. 1.5 M Tris–HCl, pH 8.8.
13. BSA solution (20 mg/mL).
14. Absorption spectrometer.

2.2 Characterization of Anti-Phosphopeptide Antibodies by Dot Blotting

1. TBS-T: TBS buffer containing 0.05% (v/v) Tween 20.
2. Amersham™-Protran™ nitrocellulose blotting membrane (0.45 μm) (Cytiva).
3. Blocking reagent: e.g., Block Ace (Yukizirushi).
4. Anti-rabbit IgG, horseradish peroxidase (HRP)-linked whole Ab donkey (e.g., Cytiva).
5. ECL reagent: SuperSignal™ West Dura Extended Duration Substrate, or SuperSignal™ West Femto (Thermo Scientific).
6. LuminoGraph III (ATTO).
7. CS Analyzer 4 to analyze chemiluminescence signals (ATTO).

2.3 Preparation of INF1 Peptide in *Escherichia coli*

1. *E. coli* strain JM109 carrying pFB53 (*infl* in pFLAG-ATS) [12].
2. LB media.
3. 100 mM IPTG.
4. Millex-HP Syringe Filter Unit, 0.45 μm, PES (Millipore).
5. Spectra/Por 3 Dialysis Sack, 3.5 kDa (Spectrum Laboratories).

2.4 *Agrobacterium*-Mediated Transient Expression in *N. benthamiana*

1. 3- to 4-week-old *N. benthamiana*.
2. *Agrobacterium tumefaciens* strain GV3101 containing 35S: FLAG-NbRBOHB or 35S:P19 (silencing suppressor) (*see* **Note 3**).
3. LB media.
4. Suspension buffer: 10 mM MES/NaOH, 10 mM $MgCl_2$, pH 5.6.
5. 150 mM acetosyringone dissolved in DMSO, aliquot, and store at −20 °C.

2.5 Preparation of Microsome Fractions from *N. benthamiana* Leaves

1. 0.5 M MOPS/KOH, pH 7.6, sterilize by filtration.
2. 1.5 M Tris–HCl, pH 7.5, sterilize by filtration.
3. 0.5 M EDTA/NaOH, pH 8.0, sterilize by filtration.
4. 0.5 M EGTA/NaOH, pH 8.0, sterilize by filtration.
5. 4 M NaCl, sterilize by filtration.
6. 1 M DTT, sterilize by filtration, aliquot, and store at −20 °C.
7. 0.5 M NaF, sterilize by filtration, aliquot, and store at −20 °C.
8. 0.1 M Na_3VO_4, sterilize by filtration, aliquot, and store at −20 °C (*see* **Note 4**).
9. 0.5 M Na_2MoO_4, sterilize by filtration, aliquot, and store at −20 °C.
10. 0.1 M AEBSF, sterilize by filtration, aliquot, and store at −20 °C.
11. 1 mM E-64 in DMSO, sterilize by filtration, aliquot, and store at −20 °C.
12. 0.1 M PMSF in 2-propanol (*see* **Note 5**).
13. Sorbitol.
14. Glycerol.
15. β-Glycerophosphate.
16. Protease inhibitor cocktail for plant cell extracts (Sigma).
17. 10% IGEPAL CA630.
18. Protein extraction buffer: 50 mM MOPS-KOH, pH 7.6, 0.5 M sorbitol, 10 mM DTT, 5 mM EGTA, 5 mM EDTA,

0.1 M NaF, 1 mM Na_3VO_4, 50 mM β-glycerophosphate, 0.1 mM AEBSF, and 1 μM E-64.

19. Suspension buffer: 150 mM Tris–HCl, pH 7.5, 50 mM NaCl, 10% glycerol, 10 mM EDTA, 10 mM DTT, 1% IGEPAL CA630, 1 mM Na_2MoO_4, 2 mM NaF, 1 mM PMSF, 1.5 mM Na_3VO_4, 1% protease inhibitor cocktail for plant cell (Sigma Aldrich), and 50 mM β-glycerophosphate.
20. Bradford assay kit (e.g., Bio-Rad Protein Assay Dye Reagent Concentrate).
21. Shake Master (Bio Medical Science).
22. Glass homogenizer with PTFE pestle.

2.6 Immunoprecipitation of RBOH Protein (Optional)

1. μMACS™ separator (Miltenyi Biotec).
2. μColumns (Miltenyi Biotec).
3. μMACS™ DYKDDDK isolation kit (Miltenyi Biotec).

2.7 Immunodetection of Phosphorylated RBOH Protein Using Anti-Phosphopeptide Antibodies

1. Standard equipment for SDS-PAGE and semi-dry transfer.
2. TBS-T: TBS buffer containing 0.05% (v/v) Tween 20.
3. Amersham™-Protran™ nitrocellulose blotting membrane (0.45 μm) (Cytiva).
4. Blocking reagent: e.g., Block Ace (Yukizirushi).
5. Anti-FLAG® M2 monoclonal antibody (Sigma).
6. Anti-rabbit IgG, HRP-linked whole Ab donkey (e.g., Cytiva).
7. Anti-mouse IgG, HRP-linked whole Ab sheep (e.g., Cytiva).
8. ECL reagent: SuperSignal™ West Dura Extended Duration Substrate, or SuperSignal™ West Femto (Thermo Scientific).
9. Chemiluminescence imager (ATTO).
10. CS Analyzer 4 to analyze chemiluminescence signals (ATTO).
11. Restore™ Western Blot Stripping Buffer (Thermo Scientific).
12. Red food coloring containing ponceau 4R (Kyoritsu-food Co., Ltd.)

3 Methods

3.1 Affinity Purification of Anti-Phosphopeptide Antibodies

1. Dissolve 3 mg of the peptides in 1 mL of coupling buffer.
2. Apply 6 mL of ice-cold 1 mM HCl onto the column.
3. Immediately apply 1 mL of the peptide solution onto the column.
4. Seal the column and incubate it for 30 min at 25 °C.
5. Apply 6 mL of buffer A onto the column.

6. Apply 6 mL of buffer B onto the column.
7. Apply 6 mL of buffer A onto the column.
8. Leave the column for 30 min at room temperature.
9. Apply 6 mL of buffer B onto the column.
10. Apply 6 mL of buffer A onto the column.
11. Apply 6 mL of buffer B onto the column.
12. Apply 2 mL of buffer C onto the column.
13. Apply 3 mL of buffer C onto the column.
14. Apply 3 mL of elution buffer onto the column.
15. Apply 10 mL of buffer C onto the column.
16. Dilute 10 mL antiserum with equal amount of buffer C and filter the mixture through a 0.45 μm filter.
17. Apply the diluted antiserum onto the column.
18. Apply 10 mL buffer C onto the column.
19. Apply 5 mL of the elution buffer onto the column and collect the fractions and immediately neutralize the eluting fractions with 1.5 M Tris–HCl, pH 8.8 (approximately 70 μL per 1 mL of eluate).
20. Add 100 μL of BSA solution (20 mg/mL) to 1 mL of the eluate and dialyze against TBS buffer at 4 °C.
21. Apply the dialyzed antibody solution onto the nonphosphopeptide-coupled column equilibrated with TBS buffer for absorption of anti-non-phosphopeptide antibodies and pass through the column twice. Collect the pass-through fraction.
22. Measure absorbance of the flow-through fraction at 280 nm. At 280 nm an absorbance of 1.0 is equivalent to an immunoglobulin concentration of 0.74 mg/mL.

3.2 Characterization of Anti-Phosphopeptide Antibodies by Dot Blotting

1. Dissolve peptides in TBS-T to final concentrations of 50 μg/mL.
2. Spot 2 μL of the peptide solutions onto a nitrocellulose membrane.
3. Dry the membrane at room temperature.
4. Block the membrane with Block Ace for 1 h at room temperature or overnight at 4 °C (*see* **Note 6**).
5. Incubate the membrane with primary antibody (0.1–10 μg/mL for purified antibodies) dissolved in TBS-T for 30 min at room temperature.
6. Wash the membrane three times with TBS-T (3 × 10 min).

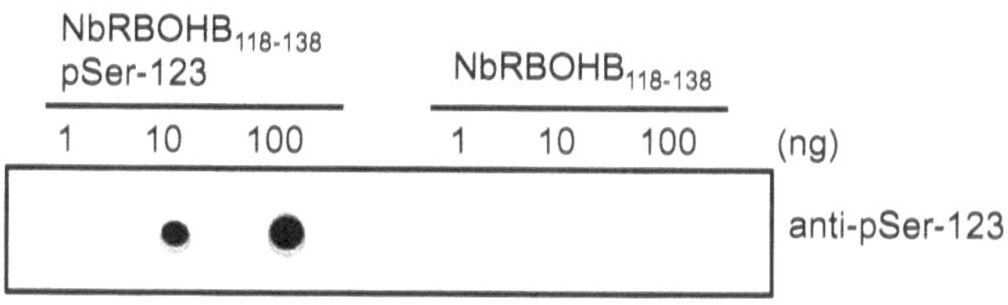

Fig. 1 Dot blot analysis of anti-phosphopeptide antibody. Non-phosphorylated peptide of NbRBOHB (NbRBOHB$_{118-138}$) and Ser-123-phosphorylated peptide (NbRBOHB$_{118-138}$pSer-123) were spotted on nitrocellulose membrane. Immunoblot analysis was performed using anti-pSer-123 antibody. The NbRBOHB$_{118-138}$pSer-123 was only detected by the antibody

7. Incubate the membrane with HRP-conjugated secondary antibody at the desired dilution for 1 h at room temperature.
8. Wash the membrane three times with TBS-T (3 × 10 min).
9. Immunoreactive signals were visualized using ECL reagent as shown in Fig. 1.

3.3 Preparation of INF1 Peptide from Escherichia coli

1. Dilute the overnight culture of *E. coli* carrying pFB53 (*infl* in pFLAG-ATS) 50-fold in LB/ampicillin.
2. Incubate the culture with shaking at 37 °C until the optical density at 600 nm (OD_{600}) reaches 0.6.
3. Apply 1 mM IPTG at final concentration to the culture to induce *infl* expression.
4. Incubate the culture with shaking for 4 h at 37 °C.
5. Centrifuge the culture at 5000 g for 10 min at 4 °C.
6. Filter the supernatant through 0.45 μm filter (PES) to remove bacterial cells.
7. Transfer the filtrated supernatant to Spectra/Por 3 Dialysis Sack, 3.5 kDa.
8. Dialyze the filtrated supernatant against 2 L of distilled water for 1 h at 4 °C.
9. Change the distilled water and dialyze the filtrated supernatant for 24 h at 4 °C.
10. Determine the protein concentration of the dialyzed suspension by the Bradford assay with BSA as a standard.
11. Adjust the protein concentration to 100 μg/mL.

3.4 Agrobacterium-*Mediated Transient Expression in* N. benthamiana

1. Transform the plasmids into *A. tumefaciens* strain GV3101 containing the transformation helper plasmid pSoup by electroporation.
2. Dilute the overnight culture threefold in LB/kanamycin/rifampicin.

3. Incubate the culture with shaking for 2–3 h at 28 °C.
4. Harvest the cells by centrifugation.
5. Resuspend the pellet in 10 mM MES-NaOH, pH 5.6, and 10 mM $MgCl_2$ using vortex.
6. Adjust the suspension to OD_{600} 0.5.
7. Mix the suspension of *A. tumefaciens* carrying 35S:FLAG-NbRBOHB and 35S:P19.
8. Add acetosyringone to a final concentration of 150 μM.
9. Incubate the bacterial suspension mixture for 1 h at room temperature.
10. Infiltrate the bacterial suspension into leaves of 4-week-old *N. benthamiana* plants using a needleless syringe.
11. Incubate *N. benthamiana* plants in growth cabinet at 23 °C for 1–3 days.

3.5 Preparation of Microsome Fractions from N. benthamiana *Leaves*

1. Infiltrate INF1 solution into *N. benthamiana* leaves 1–3 days after agroinfiltration.
2. Harvest 2–4 of *N. benthamiana* leaves without central vein.
3. Freeze *N. benthamiana* leaves with stainless beads (10 and 6 mm diameter) in falcon tube.
4. Pulverize *N. benthamiana* leaves in falcon tube by Shake Master.
5. Apply 10 mL of protein extraction buffer to the tube and immediately suspend the pulverized leaves.
6. Filter the extract through four layers of gauze and transfer the filtrate to new falcon tube.
7. Apply 2 mL of protein extraction buffer again to rinse the extract remaining in the tube and transfer the filtrate through gauze to the tube.
8. Centrifuge the filtrated extract at 10,000 g for 5 min at 4 °C.
9. Transfer the supernatant to centrifuge tube.
10. Collect 200 μL of the supernatant as total fraction.
11. Centrifuge the supernatant at 150,000 g for 40 min at 4 °C.
12. Discard the supernatant using Pasteur pipette.
13. Suspend the pellet in 1.5 mL of suspension buffer using Pasteur pipette (*see* **Note 7**).
14. Suspend the pellet in glass homogenizer with PTFE pestle and transfer the suspension to new tube.
15. Rotate the suspension gently for an hour at 4 °C to solubilize the plasma membrane.

16. Determine the protein concentration of the suspension by the Bradford assay with BSA as a standard.

3.6 Immuno-precipitation of RBOH Protein (Optional)

1. Apply 50 μL of anti-DYKDDDDK magnetic beads to the microsome fraction.
2. Rotate the microsome fraction gently for an hour at 4 °C.
3. Place μColumn in the magnetic field of μMACS separator.
4. Wash the μColumn with 500 μL of suspension buffer three times to equilibrate.
5. Apply the microsome fraction incubated with magnetic beads onto the μColumn.
6. Wash the μColumn with 500 μL of suspension buffer 3–5 times.
7. Apply 20 μL of elusion buffer (99 °C) onto the μColumn. Immediately press with finger wearing clean gloves so that the elusion buffer reaches the gray part of the μColumn and incubate for 5 min.
8. Press the μColumn with gloved finger to elute.
9. Apply 20–30 μL of elusion buffer (99 °C) onto the μColumn and repeat **steps** 7 and **8**.

3.7 Immuno-detection of Phosphorylated RBOH Protein Using Anti-Phosphopeptide Antibodies

1. Separate equal amounts of proteins on a 7.5% or 10% SDS-polyacrylamide gel (*see* **Note 8**).
2. Transfer the proteins onto a nitrocellulose membrane.
3. Block the membrane in Block Ace overnight at 4 °C.
4. Wash the membrane three times in TBS-T.
5. Incubate the membrane with anti-phosphopeptide antibody as primary antibody diluted with TBS-T at room temperature for 1 h.
6. Wash the membrane with TBS-T three times.
7. Incubate the membrane with HRP-conjugated anti-rabbit IgG antibody as a secondary antibody diluted with TBS-T for 1 h at room temperature.
8. Detect the antibody-antigen complex using ECL reagent and the chemiluminescence imager as shown in Fig. 2.
9. Wash the membrane with TBS-T three times (3 × 10 min).
10. Incubate the membrane in Restore™ Western Blot Stripping Buffer for 15–30 min.
11. Wash the membrane with TBS-T three times (3 × 10 min).
12. Repeat **steps 3–8** using anti-FLAG antibody and anti-mouse IgG antibody as primary antibody and secondary antibody, respectively.

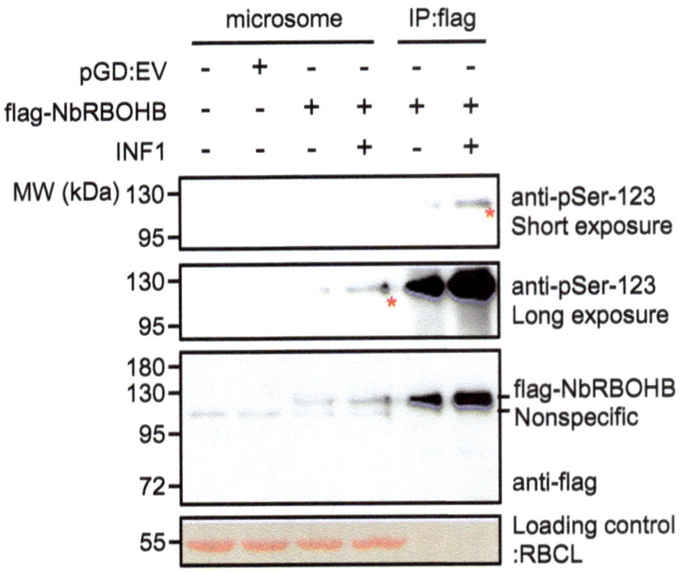

Fig. 2 In vivo detection of phospho-Ser123-NbRBOHB during immune response. FLAG-tagged NbRBOHB were expressed in *N. benthamiana* leaves by agroinfiltration. Two days after agroinfiltration, the leaves were treated with INF1 peptide. Microsomal fractions were prepared at 6 h after elicitation. After immunoprecipitation (IP) by anti-flag magnetic beads, microsome fraction (10 μg) and IP fraction (20 μL) were separated by SDS-PAGE using 10% polyacrylamide gel. Immunoblot analyses were performed using anti-NbRBOHB (pSer-123) and anti-flag antibodies. Protein loads were monitored by ponceau 4R staining of the bands corresponding to ribulose-1,5-bisphosphate carboxylase large subunit (RBCL). Red asterisks indicate the signal of anti-NbRBOHB (pSer-123) antibody increased in response to INF1

13. Wash the membrane with 1% acetic acid.
14. Incubate the membrane in 5% acetic acid with 0.3% Ponceau 4R to stain Rubisco large subunit (RBCL) as a protein loading control (*see* **Note 9**).
15. Wash the membrane with 1% acetic acid.

4 Notes

1. The P19 protein of Tomato bushy stunt virus suppresses RNA silencing. Co-expression of P19 prevents the onset of RNA silencing and allows high-level transient expression of transgenes [13].
2. Use a peptide, which consists of the phospho-serine/threonine and four or five residues on both the N- and C-terminal sides. Cysteine residue is added to the N-terminus of the peptide to

conjugate with a carrier protein, KLH (Keyhole Limpet Hemocyanin).

3. Utilize custom antibody production services, including peptide synthesis. Recently, many companies provide such services. It is also possible to produce antibodies in own laboratory following the procedure of antibody production described in the literature [14].
4. Preparation of 0.1 M Na_3VO_4 solution: Dissolve Na_3VO_4 in water and adjust to pH 10 with HCl. Boil until the solution turns colorless, then cool to room temperature and adjust the pH to 8. Repeat boil/cool/pH step until the solution remains at pH 8. Store in aliquots at −20 °C. Just before use, boil at 100 °C for 5 min, then place on ice.
5. Apply 0.1 M PMSF to the suspension solution right before use, because it is degraded after 30–60 min in an aqueous solution.
6. In the case of anti-pSer-123 NbRBOHB, we use Block Ace (Yukizirushi) as a blocking reagent, but the suitable blocking reagent needs to be selected by dot blotting. Generally, BSA is a better blocking reagent for phosphoprotein detection, because milk contains a phosphoprotein, casein, blocking with milk may produce a high background when blotting with anti-phosphopeptide antibodies.
7. The pellet is dislodged by gently scratching it with the tip of a Pasteur pipette or by using the flow of liquid. Then transfer to a glass homogenizer and grind large fragments into smaller pieces.
8. When performing immunoblotting of *N. benthamiana* protein with an anti-FLAG antibody, a non-specific band appears around 100 kDa, near the FLAG-RBOH signal (Fig. 2). These bands can be separated by adjusting the acrylamide concentration or extending the electrophoresis time. In the case of NbRBOHB, 10% SDS-polyacrylamide gel is used with a slightly longer running time.
9. Contreras et al. [15] developed Ponceau 4R staining using Irn-Bru (Scottish carbonated soft drink). We modified Ponceau 4R staining using red food coloring with Ponceau 4R as the main ingredient, because it is hard to get Irn-Bru in Japan. It is highly cost-effective compared to conventional methods using research reagents.

Acknowledgments

We thank Andrew O. Jackson for pGD binary vector, Sophien Kamoun for INF1 constructs, David C. Baulcombe for P19 construct, and the Leaf Tobacco Research Center, Japan, for

N. benthamiana seeds. This work was supported by Grant-in-Aid for Research Fellowship for Young Scientists (25KJ1409) and by Grants-in-Aid for Exploratory Research (25K22359) to from the Japan Society of the Promotion of Science.

References

1. Yuan M, Jiang Z, Bi G, Nomura K, Liu M, Wang Y, Cai B, Zhou JM, He SY, Xin XF (2021) Pattern-recognition receptors are required for NLR-mediated plant immunity. Nature 592:105–109
2. Yoshioka H, Numata N, Nakajima K, Katou S, Kawakita K, Rowland O, Jones JDG, Doke N (2003) *Nicotiana benthamiana* gp91^{phox} homologs *NbrbohA* and *NbrbohB* participate in H_2O_2 accumulation and resistance to *Phytophthora infestans*. Plant Cell 15:706–718
3. Hino Y, Inada T, Yoshioka M, Yoshioka H (2024) NADPH oxidase-mediated sulfenylation of cysteine derivatives regulates plant immunity. J Exp Bot 75:4641–4654
4. Entila F, Han X, Mine A, Schulze-Lefert P, Tsuda K (2024) Commensal lifestyle regulated by a negative feedback loop between *Arabidopsis* ROS and the bacterial T2SS. Nat Commun 15:456
5. Mittler R (2017) ROS are good. Trends Plant Sci 22:11–19
6. Castro B, Citterico M, Kimura S, Stevens DM, Wrzaczek M, Coaker G (2021) Stress-induced reactive oxygen species compartmentalization, perception and signalling. Nat Plants 7:403–412
7. Kadota Y, Sklenar J, Derbyshire P, Stransfeld L, Asai S, Ntoukakis V, Jones JD, Shirasu K, Menke F, Jones A, Zipfel C (2014) Direct regulation of the NADPH oxidase RBOHD by the PRR-associated kinase BIK1 during plant immunity. Mol Cell 54:43–55
8. Kobayashi M, Ohura I, Kawakita K, Yokota N, Fujiwara M, Shimamoto K, Doke N, Yoshioka H (2007) Calcium-dependent protein kinases regulate the production of reactive oxygen species by potato NADPH oxidase. Plant Cell 19: 1065–1080
9. Kobayashi M, Yoshioka M, Asai S, Nomura H, Kuchimura K, Mori H, Doke N, Yoshioka H (2012) StCDPK5 confers resistance to late blight pathogen but increases susceptibility to early blight pathogen in potato via reactive oxygen species burst. New Phytol 196:223–237
10. Lee D, Lal NK, Lin ZD, Ma S, Liu J, Castro B, Toruño T, Dinesh-Kumar SP, Coaker G (2020) Regulation of reactive oxygen species during plant immunity through phosphorylation and ubiquitination of RBOHD. Nat Commun 11:1838
11. Li J, Ishii T, Yoshioka M, Hino Y, Nomoto M, Tada Y, Yoshioka H, Takahashi H, Yamauchi T, Nakazono M (2025) CDPK5 and CDPK13 play key roles in acclimation to low oxygen through the control of RBOH-mediated ROS production in rice. Plant Physiol 197:kiae293
12. Kamoun S, van West P, de Jong AJ, de Groot KE, Vleeshouwers VG, Govers F (1997) A gene encoding a protein elicitor of *Phytophthora infestans* is down-regulated during infection of potato. Mol Plant-Microbe Interact 10:13–20
13. Jay F, Brioudes F, Voinnet O (2023) A contemporary reassessment of the enhanced transient expression system based on the tombusviral silencing suppressor protein P19. Plant J 113:186–204
14. Cooper HM, Paterson Y (2008) Production of polyclonal antisera. Curr Protoc Mol Biol Chapter 11:Unit 11.12
15. Contreras M, Pai H, Thompson R, Marchal C, Claeys J, Adachi H, Kamoun S (2024) The nucleotide-binding domain of NRC-dependent disease resistance proteins is sufficient to activate downstream helper NLR oligomerization and immune signaling. New Phytol 243:345–361

Chapter 8

The Application of RenSeq for Pan-NLRome Analysis

Yujing Wu, Jonathan D. G. Jones, and Xiao Lin

Abstract

Resistance gene enrichment sequencing (RenSeq) was developed in 2013. It has accelerated the cloning of plant *NLR* genes and has contributed to resistance breeding for multiple crop plants, such as potato, wheat, and rice. By combining with other strategies, many applications were developed, such as cDNA-RenSeq, dRenSeq, SMRT-RenSeq, RLP/KSeq, AgRenSeq, and MutRenSeq. These methods have been widely applied in different crops. In this protocol, we present a step-by-step guide for applying RenSeq in gene cloning and Pan-NLRome analysis. The protocol covers bait design, library preparation, target enrichment, and downstream bioinformatic analysis. This methodology can make RenSeq more accessible to researchers working with different crops and enhance our understanding of plant resistance genes in the age of pan-genome.

Key words RenSeq, *NLR* genes, Target enrichment sequencing, Pan-NLRome

1 Introduction

Plants possess a two-tier immune system. The surface immune receptors enable recognition of apoplastic molecules from pathogens, such as MAMPs and apoplastic effectors, while the NLRs (nucleotide-binding domain and leucine-rich repeat receptors) detect intracellular effectors [1]. Most cloned plant resistance genes belong to the NLR class [2]. In the early days of plant *R* gene cloning, the first two cloned *NLR* genes, *N* [3] and *RPS2* [4] were identified through transposon tagging and map-based cloning. However, developing molecular markers and generating physical maps are time- and cost-consuming. In 2013, RenSeq (Resistance gene enrichment sequencing) was developed. RenSeq represents a methodology that employs biotin-labeled RNA probes, which are complementary to *NLR* genes, to achieve targeted enrichment of *NLR* genes within a DNA library prior to sequencing. It led to the fast mapping of *Rpi-ber2* and *Rpi-rzc1* [5]. In 2016, RenSeq was updated with single molecule real time (SMRT) sequencing; this enabled assembling of all full-length

Tsutomu Kawasaki and Hirofumi Yoshioka (eds.), *Plant Immunity: Methods and Protocols*, Methods in Molecular Biology, vol. 3012, https://doi.org/10.1007/978-1-0716-5138-4_8,

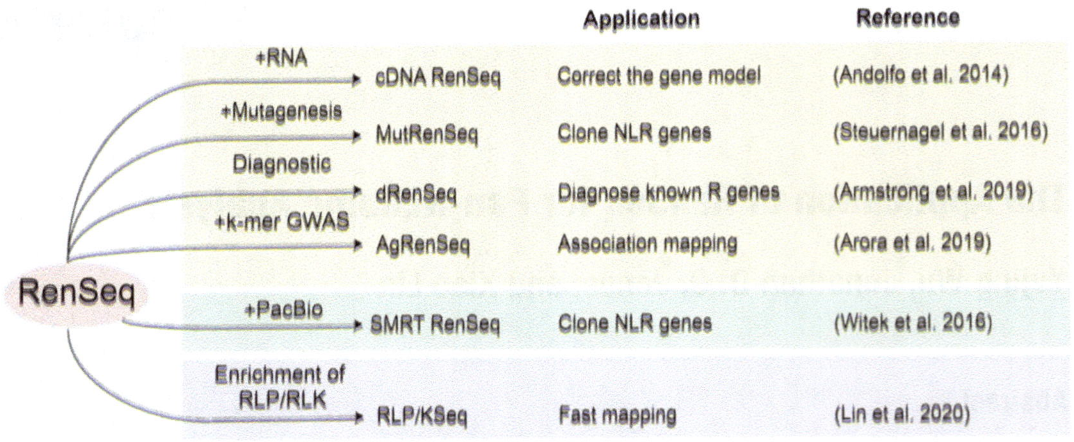

Fig. 1 Methods and applications derived from RenSeq

NLR loci, and it enabled the cloning of the late blight resistance genes *Rpi-amr3* and *Rpi-amr1* from *Solanum americanum* [6, 7].

Subsequently, a series of derivative methods have emerged, including cDNA RenSeq [8], MutRenSeq [9], dRenSeq [10], AgRenSeq [11], and RLP/KSeq [12], and led to the cloning of many crop *R* genes such as stem rust resistance genes *Sr22* and *Sr45* from hexaploid bread wheat [9]. Each of these methods has distinct applications (Fig. 1).

SMRT RenSeq enables species-level pan-NLRome analysis. In *Arabidopsis thaliana*, *NLR* sequences were enriched and sequenced using SMRT RenSeq technology across 64 ecotypes. De novo assembly and annotation were carried out, and a nearly complete pan-NLRome of *Arabidopsis thaliana* was constructed. Then, classification, diversity analysis, and selection analysis of this pan-NLRome were conducted [13]. This method was also applied to investigate the pan-NLRome in wild tomato species [14] and rice (*Oryza sativa*) [15]. SMRT RenSeq combines two key advantages. First, probe enrichment reduces genomic complexity. Second, the long read lengths of SMRT sequencing enable effective discrimination of clustered and paralogous highly similar *NLRs*. Together, these features enable the generation of a pan-NLRome at a low cost.

Here, we present a step-by-step guide for applying RenSeq in pan-NLRome analysis, covering bait design, library preparation, target enrichment, and downstream analysis. This method can be applied to any plant for which a reference NLRome is available for it or a close relative.

2 Materials

2.1 Reagents

1. VAHTS® Universal DNA Library Prep Kit for Illumina V4 (Vazyme®, ND610).
2. VAHTS® Multiplex Oligos Set 4 for Illumina (Vazyme®, N321).
3. TruePrep® DNA Library Prep Kit V2 for Illumina (Vazyme®, TD501).
4. TruePrep Index Kit V2 for Illumina (Vazyme®, TD202).
5. Nuclean Plant Genomic DNA Kit (CWBIO®, CW0531).
6. Ultrapure RNA Kit (CWBIO®, CW0581).
7. UltraClean ds-cDNA Synthesis Module (+gDNA wiper) (Vazyme®, UNR201).
8. VAHTS DNA Clean Beads (Vazyme®, N411).
9. myBaits® Target Capture Kits (Arbor Biosciences®, Cat#300316 V5).
10. Equalbit 1× dsDNA HS Assay Kit (Vazyme®, EQ121).
11. 2× Phanta Max Master Mix (Dye Plus) (Vazyme®, P525).
12. DNase/RNase-Free Water (Solarbio®, R1600), Ethanol.

2.2 Equipment

1. Centrifuge.
2. Magnetic stand.
3. Thermal cycler.
4. Metal bath.
5. 1.5 mL nuclease-free Eppendorf tube.
6. Nuclease-free PCR tube.
7. Covaris® g-TUBE (Covaris®, 50,079).
8. Qubit 4.0 Fluorometer (Thermofisher®).
9. NanoDrop.

3 Methods

3.1 Design of Biotinylated RNA-Based Probes Targeting NLR Genes

RenSeq relies on biotinylated RNA-based probes targeting genes (baits) to enrich the *NLR* genes. The bait design process involves preparing the genome of the target or a related species, predicting *NLR* genes, fragmenting them into 120 nt segments with 60 nt overlaps, and removing duplicate and paired sequences to create the bait library. This task was outsourced to Arbor Biosciences.

3.2 DNA Acquisition

3.2.1 gDNA

We used the Nuclean Plant Genomic DNA Kit (CWBIO®, CW0531) to extract high-quality DNA from approximately 100 mg of young leaves. We detected the quantity and quality of the DNA using a Nanodrop spectrophotometer and 1% agarose gel electrophoresis. It is necessary to ensure that the A260/A280 ratio is between 1.8 and 2.0, the A260/A230 ratio is greater than 2.0, and a single clear band appears in the agarose gel electrophoresis.

3.2.2 cDNA

We used the Ultrapure RNA Kit (CWBIO®, CW0581) to extract RNA from approximately 50 mg of young leaves. The RNA concentration was measured using a Nanodrop. Then, 1 μg of the RNA was used to synthesize the double-stranded cDNA with the Ultra-Clean ds-cDNA Synthesis Module (with gDNA wiper) (Vazyme®, UNR201).

3.3 Long-Fragment Library Construction (>2000 bp)

3.3.1 DNA Fragmentation

We used the Covaris® g-TUBE to fragment DNA into approximately 5 kb-sized pieces. Detailed steps are as follows:

1. Dilute the DNA (>3 μg) to a final volume of 160 μL with nuclease-free water.
2. Unscrew the g-TUBE cap and add 150 μL of the DNA to the tube top and tighten the tube cap (*see* **Note 1**). Immediately perform centrifugation at 13,800× *g* for 30 s.
3. Invert the g-TUBE (screw-cap facing downwards), and then centrifuge again at 13,800× *g* for 30 s.
4. Remove the g-TUBE from the centrifuge while keeping it inverted. Unscrew the cap and transfer the solution in the cap to a nuclease-free 1.5 mL Eppendorf tube. In this way, the fragmented DNA sample is obtained.
5. Take 5 μL of the fragmented DNA sample, run it on a 1% agarose gel, then check the size range.

3.3.2 Purification and Size Selection of Sheared DNA

1. Allow the Vazyme® DNA clean beads (N411) to equilibrate at room temperature for 30 min to ensure they reach ambient temperature.
2. Transfer 150 μL of sheared DNA to a 1.5 mL nuclease-free Eppendorf tube and add 60 μL (0.4×) DNA clean beads (pipette after mixing).
3. Mix by pipetting up and down and incubate at room temperature for 5 min.
4. Centrifuge briefly and place the tube on a magnetic stand for 5 min to allow the magnetic beads and the solution to separate, then aspirate off the supernatant.
5. Add 200 μL of freshly prepared 80% ethanol to the tube while on the magnetic stand. Incubate at room temperature for 30 s, and then carefully remove and discard the supernatant. Repeat the wash twice for a total of three washes.

6. Centrifugate briefly and return the tube to the magnetic stand. Use a small-volume (10 μL) pipette to remove the remaining liquid. Then, keep the tube on the magnetic stand with the lid open and air-dry the beads at room temperature for 10 min.
7. Remove the tube from the magnetic stand, add 50 μL of nuclease-free water, mix thoroughly with a pipette, and then let it stand at room temperature for 2 min. After that, return the tube to the magnetic stand until the solution is clear (about 5 min), and transfer the supernatant to a new nuclease-free tube (*see* **Note 2**).

3.3.3 End Preparation

The amount of input DNA should be within the range of 100 ng to 1 μg.

1. Prepare the following reaction mixture in a nuclease-free PCR tube.

End prep enzyme 2	3 μL
End prep buffer 2	7 μL
Fragmented DNA	50 μL
Total volume	60 μL

2. Vortex to mix thoroughly and then perform a brief centrifugation to collect the solution at the bottom of the tube.
3. Place the tube into the thermal cycler and run the following program.

20 °C	30 min
65 °C	30 min
4 °C	∞
Lid 105 °C	

3.3.4 Adaptor Ligation

1. Prepare the following reaction mixture in the PCR tube from the previous step.

End preparation reaction mixture	60 μL
Rapid DNA ligase 2	30 μL
Rapid ligation buffer 5	10 μL
DM5xx[a]	2.5 μL
DM7xx[a]	2.5 μL
ddH_2O	5 μL
Total volume	110 μL

[a]From VAHTS® Multiplex Oligos Set 4 for Illumina (Vazyme®, N321) (*see* **Note 3**)

2. Vortex to mix thoroughly and then perform a brief centrifugation to collect the solution at the bottom of the tube.
3. Place the tube into the thermal cycler and run the following program.

20 °C	15 min
4 °C	∞
Lid 105 °C	

3.3.5 Cleanup of Adapter-Ligated DNA

Purify DNA from the ligation reaction mixture with 55 μL (0.5×) DNA clean beads as described in Subheading 3.3.1, **step 2**, and elute it in 50 μL nuclease-free water.

3.3.6 PCR Amplification

1. Prepare the following reaction mixture in a nuclease-free PCR tube.

Purified adapter ligation product	20 μL
VAHTS HiFi Amplification Mix 2	25 μL
PCR Primer Mix 3 for Illumina	5 μL
Total volume	50 μL

2. Vortex to mix thoroughly and then perform a brief centrifugation to collect the solution at the bottom of the tube.
3. Place the tube into the thermal cycler and run the following program. Repeat ii–iv for 5–20 cycles. Initially, carry out 5 cycles, and then load 5 μL of the reaction mixture onto a 1% agarose gel alongside 1 μL of unamplified library DNA (serving as a control for the input DNA). If the DNA yield is insufficient, return the reaction to the thermocycler for an extra 5 cycles (*see* **Note 4**).

i	98 °C	30 s
ii	98 °C	30 s
iii	65 °C	30 s
iv	72 °C	4 min
v	72 °C	10 min
vi	4 °C	∞
Lid 105 °C		

3.3.7 Cleanup of PCR-Amplified DNA

Purify DNA from the 30 μL PCR mixture with 15 μL (0.5×) DNA clean beads as described in Subheading 3.3.1, **step 2**, and elute it in 5–6 μL nuclease-free water.

3.4 Short-Fragment Library Construction (<500 bp)

We successfully used the TruePrep® DNA Library Prep Kit V2 for Illumina (Vazyme®, TD501), a kit based on the activity of TN5 transposase, to rapidly construct short-fragment libraries. We used 50 ng of DNA for library construction.

3.4.1 DNA Fragmentation and Adaptor Ligation

1. Prepare the following reaction mixture in a PCR tube.

DNA	1 μL
TTE Mix V50	2.5 μL
5×TTBL	5 μL
ddH_2O	16.5 μL
Total volume	25 μL

2. Vortex to mix thoroughly and then perform a brief centrifugation to collect the solution at the bottom of the tube.
3. Place the tube into the thermal cycler and run the following program.

55 °C	10 min
10 °C	∞
Lid 105 °C	

3.4.2 Purification of Sheared DNA

1. Allow the DNA clean beads to equilibrate at room temperature for 30 min to ensure they reach ambient temperature.
2. Add 25 μL (1×) DNA clean beads (pipette after mixing) to the previous reaction mixture.
3. Mix by pipetting up and down and incubate at room temperature for 5 min.
4. Centrifugate briefly and place the tube on a magnetic stand for 5 min to allow the magnetic beads and the solution to separate, then aspirate off the supernatant.
5. Add 200 μL of freshly prepared 80% ethanol to the tube while on the magnetic stand. Incubate at room temperature for 30 s, and then carefully remove and discard the supernatant. Repeat the wash twice for a total of three washes.
6. Centrifuge briefly and return the tube to the magnetic stand. Use a small-volume (10 μL) pipette to remove the remaining

liquid. Then, keep the tube on the magnetic stand with the lid open and air-dry the beads at room temperature for 10 min.

7. Remove the tube from the magnetic stand, add 12 μL of nuclease-free water, mix thoroughly with a pipette, and then let it stand at room temperature for 2 min. After that, return the tube to the magnetic stand until the solution is clear (about 5 min), and transfer the supernatant to a new nuclease-free tube.

3.4.3 PCR Amplification

1. Prepare the following reaction mixture in a nuclease-free PCR tube.

Purified adapter ligation product	12 μL
TAE	0.5 μL
PPM	2.5 μL
N5xx[a]	2.5 μL
N7xx[a]	2.5 μL
5×TAB	5 μL
Total volume	25 μL

[a]From the TruePrep Index Kit V2 for Illumina (Vazyme®, TD202)

2. Vortex to mix thoroughly and then perform a brief centrifugation to collect the solution at the bottom of the tube.

3. Place the tube into the thermal cycler and run the following program. Repeat iii–v for 5–15 cycles. Initially, carry out 5 cycles, and then load 5 μL of the reaction mixture onto a 1% agarose gel alongside 1 μL of unamplified library DNA (serving as a control for the input DNA). If the DNA yield is insufficient, return the reaction to the thermocycler for an extra 5 cycles.

i	72 °C	3 min
ii	98 °C	30 s
iii	98 °C	15 s
iv	60 °C	30 s
v	72 °C	3 min
vi	72 °C	5 min
vii	4 °C	∞
Lid 105 °C		

3.4.4 Size Selection

We screened DNA fragments by size through two rounds of purification using DNA clean beads at different concentrations. Here, we screened fragments around 350 bp after adapter ligation. Alternatively, longer fragments can be screened as instructed in the manual of the TruePrep® DNA Library Prep Kit V2 for Illumina (Vazyme®, TD501).

1. Make up the volume of the PCR amplification mixture to 25 μL with nuclease-free water.
2. Add 17.5 μL (0.75×) DNA clean beads (equilibrated to room temperature; pipette after mixing) to the PCR amplification mixture.
3. Mix by pipetting up and down and incubate at room temperature for 5 min.
4. Centrifuge briefly and place the tube on a magnetic stand for 5 min to allow the magnetic beads and the solution to separate. Then transfer the supernatant to a new nuclease-free tube and discard the beads.
5. Add 3.76 μL (0.15×) DNA clean beads (equilibrated to room temperature; pipette after mixing) to the above-mentioned supernatant.
6. Mix by pipetting up and down and incubate at room temperature for 5 min.
7. Centrifuge briefly and place the tube on a magnetic stand for 5 min to allow the magnetic beads and the solution to separate. Aspirate and discard the supernatant.
8. Add 200 μL of freshly prepared 80% ethanol to the tube while on the magnetic stand. Incubate at room temperature for 30 s, and then carefully remove and discard the supernatant. Repeat wash twice, to three washes in total.
9. Centrifugate briefly and return the tube to the magnetic stand. Use a small-volume (10 μL) pipette to remove the remaining liquid. Then, keep the tube on the magnetic stand with the lid open and air-dry the beads at room temperature for 10 min.
10. Remove the tube from the magnetic stand, add 5–6 μL of nuclease-free water, mix thoroughly with a pipette, and then let it stand at room temperature for 2 min. After that, return the tube to the magnetic stand until the solution is clear (about 5 min), and transfer the supernatant to a new nuclease-free tube.

3.5 Target Enrichment

Here, we used the myBaits® Target Capture Kit customized from Arbor Biosciences. The bait-target sequence divergence was expected to be 15–25%. The amount of DNA we added in this step ranged from 250 to 500 ng.

3.5.1 Hybridization

1. Thaw the following solutions in advance and then add them to a nuclease-free PCR tube in sequence. Vortex to mix thoroughly and then perform a brief centrifugation to collect the solution at the bottom of the tube.

HybN (Box#1)	4.64 μL
HybD (Box#2)	1.75 μL
HybS (Box#1)	0.25 μL
HybR (Box#2)	0.625 μL
HybO (Box#2)	2.5 μL
HybX (Box#2)	0.25 μL
Baits (Box#3)[a]	3 μL
Total volume	13.015 μL

[a]After first dissolution, aliquot into nuclease-free, low-adsorption tubes to minimize freeze-thaw cycles

2. Add 3–5 μL of the final product from Step 3.3 to the above tube. Mix by pipetting up and down.
3. Place the tube into the thermal cycler and run the following program. Between steps ii and iii, the temperature decrease from 95 °C to 60 °C should be carried out at a rate of 0.1 °C/s.

i	60 °C	10 min
ii	95 °C	10 min
iii	60 °C	1 min
Lid 105 °C		

4. Immediately afterwards, continue to incubate in the thermal cycler at 60 °C for 16–24 h with the heated lid at 70 °C.

3.5.2 Preparation of Streptavidin-Coated Magnetic Beads

1. Pipette 15 μL of streptavidin-coated magnetic beads (Box#1) into a nuclease-free 1.5 mL Eppendorf tube after mixing. Place the tube on a magnetic rack and let it stand for 2 min until the suspension becomes clear. While keeping the tube on the magnetic rack, aspirate and discard the supernatant.
2. Remove the tube from the magnetic rack and add 100 μL of binding buffer (Box#1) to wash the beads. Vortex to mix well and then briefly centrifuge. Place the tube back on the magnetic rack and let it stand for 2 min until the solution becomes clear. Then aspirate and discard the supernatant. Repeat this process twice for a total of three washes.

3. Add 35 μL of binding buffer (Box#1) to resuspend the beads.
4. Place the prepared beads in a heat block at 60 °C.

3.5.3 Binding Streptavidin Beads

Add the DNA library hybridized with baits (Subheading 3.4.1) to the streptavidin beads pre-heated to 60 °C (Subheading 3.4.2). Pipette up and down to mix well, and then place it on a heat block at 60 °C for incubation for 45 min. During this period, shake and mix the beads every 5–10 min.

3.5.4 Bead Washing

1. Prepare Washing Buffer X according to the following proportions, and then heat it to 60 °C using a heat block.

HybS (Box#1)	8 μL
Wash Buffer (Box#1)	200 μL
ddH_2O	792 μL
Total volume	1 mL

2. Place the tube from step 3.4.3 on a magnetic rack and let it stand for 2 min to allow the magnetic beads to be attracted. Then aspirate and discard the supernatant.
3. Add 100 μL of Wash Buffer X (preheated to 60 °C) to resuspend the beads and pipette up and down to mix well. Incubate at 60 °C for 5 min, and shake at the 2.5 min mark to mix. Then place the tube back on the magnetic rack for 2 min to allow the magnetic beads to be attracted, and aspirate and discard the supernatant. Repeat this process twice for a total of three washes.
4. Centrifuge briefly and return the tube to the magnetic stand. Use a small-volume (10 μL) pipette to remove the remaining liquid.
5. Resuspend the magnetic beads in 10–20 μL of Buffer E (Box#2).

3.6 Amplification and Purification of the Captured DNA Library

1. Prepare the following reaction mixture in a nuclease-free PCR tube.

Step 3.4 product (with beads)	10 μL
2× Phanta Master ix	25 μL
Primer F[a]	2.5 μL
Primer R[a]	2.5 μL
ddH_2O	10 μL
Total volume	50 μL

[a]Design based on the sequence information of the ligated adaptor

2. Vortex to mix thoroughly and then perform a brief centrifugation to collect the solution at the bottom of the tube.
3. Place the tube into the thermal cycler and run the following program. Repeat ii–iv for 25 cycles. Initially, carry out 5 cycles, and then load 5 μL of the reaction mixture onto a 1% agarose gel alongside 1 μL of unamplified library DNA (serving as a control for the input DNA). If the DNA yield is insufficient, return the reaction to the thermocycler for an extra 5 cycles.

i	98 °C	4 min
ii	98 °C	30 s
iii	60 °C	30 s
iv	72 °C	4 min (long fragment)/1 min (short fragment)
v	72 °C	10 min
vi	4 °C	∞
Lid 105 °C		

4. Purify DNA from the 50 μL PCR mixture with 20 μL (0.4×, long fragment)/50 μL (1×, short fragment) DNA clean beads as described in Subheading 3.3.1, **step 2**, and elute it in 30 μL nuclease-free water.

3.7 Library Quality Control

3.7.1 Library Quantification

We used the Qubit 4.0 Fluorometer to conduct a preliminary quantification of the captured DNA library.

1. Dispense 2 mL Equalbit 1× dsDNA HS Working Solution into a brown centrifuge tube and then equilibrate it to room temperature. Meanwhile, equilibrate Equalbit 1× dsDNA HS Standard #1 (0 ng/μL in TE buffer) and Equalbit 1× dsDNA HS Standard #2 (10 ng/μL in TE buffer) to room temperature.
2. Preparation of standard samples: take 190 μL of the Working Solution into two 0.5 mL transparent Eppendorf tubes, respectively. Add 10 μL of Standard #1 and Standard #2 into the two tubes, respectively. After gently vortexing to mix well, briefly centrifuge the tubes, and avoid generating bubbles during the process. Ensure accurate pipetting throughout the procedure.
3. Preparation of test samples: Pipette 199 μL of the Working Solution into separate 0.5 mL transparent Eppendorf tubes. Then, add 1 μL of the captured DNA library to be tested into the corresponding tubes. Gently vortex the tubes to mix the contents, followed by a brief centrifugation. Avoid creating air bubbles during this process. Ensure accurate pipetting throughout the procedure.

4. Incubate the prepared samples at room temperature in the dark for at least 2 min.
5. Turn on the Qubit 4.0 Fluorometer, select "dsDNA > dsDNA High Sensitivity > Read standards", and then place Standard Sample #1 and #2 into the instrument one by one as prompted by the device for detection to draw a standard curve.
6. Place the samples to be tested into the fluorometer, click "Run samples." Then adjust "the sample volume added to the assay tube" to 1 μL, select "ng/μL" as the "output sample units," and then click "Read tube" to read the values.
7. If the instrument indicates that the values exceed the measurement range, you can dilute the samples tenfold and then conduct the detection again.

3.7.2 Library Fragment Size Range

Take 5 μL of the captured DNA library and run it on a 1% agarose gel. Then, compare the bands with a suitable DNA marker to see if the DNA fragments are in the expected size range (*see* **Note 5**).

3.7.3 Enrichment Level Detection

We roughly characterized the target enrichment efficiency by detecting the abundance differences of some conserved NB-LRR sequences relative to 18S in the captured DNA library and the initial input DNA using qRT-PCR.

The information on the primer sequences we used is shown in the following table. The reagent used is Taq Pro Universal SYBR qPCR Master Mix (Vazyme®, Q712). For the specific experimental system and procedure, please refer to the reagent instruction manual.

Primer name	Sequence 5′-3′
NRC3 F	GAGAAGAAGGTTTTTCACAAGGAAAAGTGTC
NRC3 R	GTCTTTCCTTTGCCTATCAAAGCTC
NRC2b F	GCTTTCAAATCCCTGCTTGGAAG
NRC2b R	TCCATTACCATCACTAGATTCCTGTTG
18S F	AACTTAAAGGAATTGACGGAAGG
18S R	AAGTTTCCCCGTGTTGAGTC

3.8 Assembly and Analysis of the Pan-NLRome

Using gDNA as the template, we carried out long-read library construction followed by SMRT sequencing. Fragments in the range of 2 kb–5 kb are screened, and the obtained data can be de novo assembled by Canu [16] to obtain the pan-NLRome. Then, we can conduct clustering, diversity analysis, evolutionary analysis, and other analyses of *NLR* genes, with the code reference available at https://github.com/weigelworld/pan-nlrome/ [13].

If there is cDNA RenSeq data available, it can be used as a reference to correct the pan-NLRome.

In the case of an existing reference pan-NLRome, short-read RenSeq can be used to identify mutation sites in *NLR* genes. Specifically, the reads from Illumina sequencing can be directly mapped to the reference pan-NLRome using the Geneious software, and then the mutations in the reads mapped to *NLRs* can be directly examined.

4 Notes

1. If the volume of the solution added to the g-TUBE exceeds 150 μL, it will cause structural damage to the g-TUBE during the centrifugation process.
2. In this protocol, for unspecified quantification steps, Qubit quantification is recommended.
3. Ensure that the adaptors used do not have sequences complementary to the baits.
4. During the PCR amplification process, there will be sequence preference and short fragments will be enriched. Please avoid using an excessively high number of cycles.
5. Analysis of the library fragment size range can also be performed using the Agilent 2100 Bioanalyzer.

References

1. Jones JDG, Staskawicz BJ, Dangl JL (2024) The plant immune system: from discovery to deployment. Cell 187:2095–2116
2. Kourelis J, Van Der Hoorn RAL (2018) Defended to the nines: 25 years of resistance gene cloning identifies nine mechanisms for R protein function[OPEN]. Plant Cell 30:285–299
3. Whitham S, Dinesh-Kumar SP, Choi D et al (1994) The product of the tobacco mosaic virus resistance gene *N*: similarity to toll and the interleukin-1 receptor. Cell 78:1101–1115
4. Bent AF, Kunkel BN, Dahlbeck D et al (1994) RPS2 of Arabidopsis thaliana: a leucine-rich repeat class of plant disease resistance genes. Science 265:1856–1860
5. Jupe F, Witek K, Verweij W et al (2013) Resistance gene enrichment sequencing (RenSeq) enables reannotation of the NB-LRR gene family from sequenced plant genomes and rapid mapping of resistance loci in segregating populations. Plant J Cell Mol Biol 76:530–544
6. Witek K, Lin X, Karki HS et al (2021) A complex resistance locus in Solanum americanum recognizes a conserved Phytophthora effector. Nat Plants 7:198–208
7. Witek K, Jupe F, Witek AI et al (2016) Accelerated cloning of a potato late blight–resistance gene using RenSeq and SMRT sequencing. Nat Biotechnol 34:656–660
8. Andolfo G, Jupe F, Witek K et al (2014) Defining the full tomato NB-LRR resistance gene repertoire using genomic and cDNA RenSeq. BMC Plant Biol 14:120
9. Steuernagel B, Periyannan SK, Hernández-Pinzón I et al (2016) Rapid cloning of disease-resistance genes in plants using mutagenesis and sequence capture. Nat Biotechnol 34:652–655
10. Armstrong MR, Vossen J, Lim TY et al (2019) Tracking disease resistance deployment in potato breeding by enrichment sequencing. Plant Biotechnol J 17:540–549
11. Arora S, Steuernagel B, Gaurav K et al (2019) Resistance gene cloning from a wild crop

relative by sequence capture and association genetics. Nat Biotechnol 37:139–143

12. Lin X, Armstrong M, Baker K et al (2020) RLP/K enrichment sequencing; a novel method to identify receptor-like protein (RLP) and receptor-like kinase (RLK) genes. New Phytol 227:1264–1276
13. Van de Weyer A-L, Monteiro F, Furzer OJ et al (2019) A species-wide inventory of NLR genes and alleles in *Arabidopsis thaliana*. Cell 178: 1260–1272.e14
14. Seong K, Seo E, Witek K et al (2020) Evolution of NLR resistance genes with noncanonical N-terminal domains in wild tomato species. New Phytol 227:1530–1543
15. Gladieux P, van Oosterhout C, Fairhead S et al (2024) Extensive immune receptor repertoire diversity in disease-resistant rice landraces. Curr Biol 34:3983–3995.e6
16. Koren S, Walenz BP, Berlin K et al (2017) Canu: scalable and accurate long-read assembly via adaptive k-mer weighting and repeat separation. Genome Res 27:722–736

Chapter 9

Blue Native-PAGE Assay of NLR Protein Complexes Combined with Genetic Complementation in *Nicotiana benthamiana*

Kodai Honda, Tadashi Fujiwara, and Hiroaki Adachi

Abstract

BN-PAGE (Blue Native-Polyacrylamide Gel Electrophoresis) is a non-denaturing electrophoretic technique used to analyze the molecular weight and oligomeric states of protein complexes under near-native conditions. NLRs (nucleotide-binding leucine-rich repeat proteins), which function as intracellular immune receptors in plants, form oligomeric higher-order complexes known as resistosomes upon activation by recognition of pathogen effectors—a mechanism elucidated through BN-PAGE and structural analyses. Here, we describe a method combining BN-PAGE with *Agrobacterium*-mediated complementation assay to investigate the resistosome formation of the NLR protein ZAR1 in *Nicotiana benthamiana*.

Key words BN-PAGE, Complementation assay, Agroinfiltration, Western blotting, NLR, Protein complex, *Nicotiana benthamiana*

1 Introduction

BN-PAGE (Blue Native-Polyacrylamide Gel Electrophoresis) is a powerful technique for analyzing the molecular weight and oligomeric states of protein complexes. Unlike SDS-PAGE (Sodium Dodecyl Sulfate-Polyacrylamide Gel Electrophoresis), which denatures proteins using SDS, reducing agents, and heat, BN-PAGE preserves protein complexes in their native states. This is achieved by using a mild, negatively charged dye, Coomassie Brilliant Blue G-250 (CBB G-250), instead of SDS, and by omitting heat treatment to prevent protein denaturation [1–3]. As a result, proteins retain their native conformations and are separated based on their molecular masses. Furthermore, the use of mild detergents such as digitonin and *n*-dodecyl-β-D-maltoside (DDM) enables the analysis of membrane proteins under near-physiological conditions [1].

Nicothiana benthamiana is widely used as a model plant system in plant immunity research, due to its high susceptibility to a wide

Tsutomu Kawasaki and Hirofumi Yoshioka (eds.), *Plant Immunity: Methods and Protocols*, Methods in Molecular Biology, vol. 3012,
https://doi.org/10.1007/978-1-0716-5138-4_9,

variety of pathogens, which facilitates analyzing plant-pathogen interactions, and its high efficiency of transient gene expression via the agroinfiltration method [4, 5]. Agroinfiltration can be applied to virus-induced gene silencing (VIGS), RNA interference using hairpin RNAs, cell death assay, biochemical and cell biological analyses, as well as structural analyses [4–6]. Notably, these approaches can also be combined with complementation assays, in which hairpin RNA targeting the gene of interest is co-expressed with a codon-shuffled version of the corresponding protein, thereby eliminating the influence of endogenous proteins [7]. This strategy offers a simpler and faster alternative to complementation assays using transgenic lines.

NLRs (Nucleotide-binding leucine-rich repeat proteins) play a central role in plant immunity by recognizing pathogen effectors and inducing hypersensitive cell death and various defense responses [8]. Although the molecular mechanisms underlying NLR-mediated immune activation remained elusive for many years, breakthrough studies revealed that *Arabidopsis thaliana* NLR ZAR1 (HOPZ-ACTIVATED RESISTANCE1) forms an oligomeric higher-order complex called resistosome upon effector recognition through its partner receptor-like cytoplasmic kinases (RLCKs) [9, 10]. The ZAR1 resistosome functions as a calcium ion channel at the plasma membrane [11]. Subsequently, various plant NLRs have been shown to form resistosomes upon activation, as demonstrated through BN-PAGE and structural analyses [12–19]. These findings suggest that resistosome formation is a conserved and fundamental mechanism of immune activation among plant NLRs.

ZAR1 gene is an atypically conserved NLR gene among flowering plant species [20, 21]. In *N. benthamiana*, the ZAR1 ortholog (NbZAR1) recognizes the bacterial effector XopJ4 from *Xanthomonas perforans* via its partner RLCK, NbJIM2, and triggers the hypersensitive cell death response [22]. Here, we describe a detailed protocol for analyzing NbZAR1 resistosome formation by combining *Agrobacterium*-mediated complementation assay and BN-PAGE.

2 Materials

2.1 Agroinfiltration for RNA Interference and Gene Complementation

1. *Agrobacterium tumefacience* GV3101 strains transformed with gene expression vectors (*see* Tables 1, 2, and **Note 1**).
2. Four-week-old *Nicotiana benthamiana* grown at 22–26 °C, 40–60% relative humidity, under a 16-h/8-h light/dark cycle (*see* **Note 2**).
3. Luria-Bertani (LB) agar plates: 2 tablets of LB Broth (Sigma, L7275), 1.5 g agar, 100 mL pure water, 50 μg/mL kanamycin,

Table 1
Plasmid constructs used in Fig. 1

Vector	Promoter	Coding sequence	C-Tag	Terminator	OD_{600}	Reference
pRNAi-GG		hpRNA: GUS			0.5	[7]
pRNAi-GG		hpRNA: NbZAR1			0.5	[7]
pICH86988	35S	$NbZAR1^{WT}$	6xHA	OCS	0.5	[7]
pICH86988	35S	$NbZAR1^{syn}$	6xHA	OCS	0.5	[7]
pICH86988	35S	$NbZAR1^{L17E}$	6xHA	OCS	0.5	[7]
pICH86988	35S	$NbZAR1^{syn-L17E}$	6xHA	OCS	0.5	[7]

Table 2
Plasmid constructs used in Fig. 3

Vector	Promoter	Coding sequence	C-Tag	Terminator	OD_{600}	Reference
pRNAi-GG		hpRNA: NbZAR1			0.5	[7]
pICH86988	35S	$NbZAR1^{syn-L17E}$	6xHA	OCS	0.5	[7]
pICH86988	35S	NbJIM2	3xFLAG	OCS	0.1	[7]
pICH86988	35S	XopJ4		OCS	0.2	[7]
pICH86988	35S	EV		OCS	0.2	Add gene no. 48076

and 50 μg/mL rifampicin; autoclaved at 121 °C for 20 min and stored at 4 °C. Add antibiotics after autoclaving: 1/1000 volume of 50 mg/mL kanamycin (dissolved in pure water and store at −20 °C) and 50 mg/mL rifampicin (dissolved in dimethyl sulfoxide (DMSO) and stored at −20 °C).

4. MES buffer: 10 mM MES, 10 mM $MgCl_2$, adjust pH to 5.6 with NaOH; autoclaved at 121 °C for 20 min and stored at room temperature.
5. Agroinfiltration buffer: MES buffer supplemented with 150 μM acetosyringone [add 1/1000 volume of 150 mM acetosyringone (Apollo Scientific, BIA2203), dissolved in DMSO and stored at −20 °C) to the MES buffer just before use].
6. 1 mL needleless syringe.

2.2 Protein Extraction

1. Six leaf disks (8 mm diameter each) from *Nicotiana benthamiana* leaves (*see* **Note 3**).
2. Zirconia beads (HIRA CERAMICS).
3. ShakeMaster® AUTO (Bio Medical Science Inc.).

4. Protein extraction buffer: 50 mM Tris–HCl (pH 7.5), 50 mM NaCl, 5 mM $MgCl_2$, 10% (v/v) glycerol, 10 mM dithiothreitol (DTT), 1× protease inhibitor cocktail (Sigma, P9599), and detergent [select an appropriate detergent depending on the properties of the proteins. we used 0.2% or 1% (w/v) digitonin (Sigma, D141) and 0.2% or 1% (w/v) NP-40 (Thermo Scientific, 85,124)]; stored at 4 °C (*see* **Notes 4** and **5**). Add DTT and protease inhibitor cocktail just before protein extraction.
5. NativePAGE™ 5% G-250 Sample Additive (Invitrogen, BN2004); stored at room temperature.
6. NativePAGE™ 4× Sample buffer (Invitrogen, BN2003); stored at 4 °C.

2.3 BN-PAGE

1. XCell *SureLock*™ Mini-Cell (Invitrogen).
2. Native PAGE anode buffer: Dilute 50 mL of NativePAGE™ (20×) Running Buffer (Invitrogen, BN2001) with pure water to a final volume of 1 L; stored at 4 °C (*see* **Note 5**).
3. Dark cathode buffer: Mix 10 mL of NativePAGE™ (20×) Running Buffer and 10 mL NativePAGE™ (20×) Cathode Buffer Additive (Invitrogen, BN2002), and dilute with pure water to a final volume of 200 mL; stored at 4 °C (*see* **Note 5**).
4. Light cathode buffer: Mix 10 mL of NativePAGE™ (20×) Running Buffer and 1 mL of NativePAGE™ (20×) Cathode Buffer Additive, and dilute with pure water to a final volume of 200 mL; stored at 4 °C (*see* **Note 5**).
5. NativePAGE™ 3–12% Bis-Tris Mini Protein Gels (Invitrogen, BN1001 BOX); stored at 4 °C.
6. NativeMark™ Unstained Protein Standard (Invitrogen, 2347286); stored at −20 °C.

2.4 SDS-PAGE

1. Mini-PROTEAN® Tetra Vertical Electrophoresis Cell (Bio-Rad).
2. SDS-PAGE running buffer: 25 mM Tris, 192 mM glycine, 0.1% (w/v) SDS; stored at room temperature.
3. 2× SDS-PAGE loading dye: Mix 500 μL of sample Buffer Solution (2ME-) (×4) (Wako, 198-13282) with 100 μL of 1 M DTT, and dilute with pure water to a final volume of 1 mL.
4. Mini-PROTEAN® TGX™ Precast Gels (Bio-Rad, 4561095); stored at 4 °C.
5. Precision Plus Protein Dual Color Standards (Bio-Rad, 1610374); stored at 4 °C.

2.5 Western Blotting

1. Trans-Blot® Turbo™ Transfer System (Bio-Rad).
2. Trans-Blot® Turbo™ RTA Mini 0.2 μm PVDF Transfer Kit (Bio-Rad, 1704272): includes filter paper and PVDF membrane.
3. Transfer buffer for BN-PAGE: Mix 50 mL of NuPAGE™ (20×) Transfer Buffer (Invitrogen, NP0006) with 100 mL of 100% (v/v) methanol, and dilute with pure water to a final volume of 1 L. Store at 4 °C (*see* **Note 5**).
4. Transfer buffer for SDS-PAGE: Mix 200 mL of Trans-Blot Turbo™ 5× Transfer Buffer (Bio-Rad, 10026938) with 200 mL of 100% (v/v) ethanol, and dilute with pure water to a final volume of 1 L. Store at 4 °C (*see* **Note 5**).
5. 100% (v/v) methanol.
6. 8% (v/v) acetic acid.
7. 100% (v/v) ethanol.
8. SDS solution: 50 mM Tris–HCl (pH 7.5), 2% (w/v) SDS, 150 mM DTT; stored at room temperature. Add DTT before use.
9. Tris-buffered saline with Tween (TBS-T): 50 mM Tris–HCl (pH 7.4), 137 mM NaCl, 2.68 mM KCl, 0.05% (v/v) TWEEN® 20 (Sigma, P9416); stored at room temperature.
10. Blocking solution: 3% (w/v) skim milk powder in TBS-T; prepare just before use.
11. Antibody: anti-HA-Peroxidase (Merck, 12013819001); stored at −20 °C.
12. SuperSignal™ West Dura Extended Duration Substrate (Thermo Scientific, 34075).
13. Imaging film (azonx, AZ-550).
14. LuminoGraph II EM (ATTO).
15. Ponceau-S staining solution (BEA, BCL-PSS-01); stored at room temperature.

3 Methods

3.1 RNA Interference of Endogenous NLR Gene

1. Spread *Agrobacterium* strain carrying the binary vector for expressing hairpin RNA targeting endogenous NbZAR1 (hpRNA:NbZAR1) onto LB agar plates containing 50 μg/mL kanamycin and 50 μg/mL rifampicin. Incubate at 28 °C for 1–2 days.
2. Collect the cultured *Agrobacterium* strain and resuspend in 1 mL MES buffer.
3. Centrifuge at 13,000× *g* for 2 min and discard the supernatant.

4. Resuspend the pellet in 1 mL agroinfiltration buffer and incubate at room temperature for 2 h.
5. Measure OD_{600} and adjust to $OD_{600} = 0.5$ with agroinfiltration buffer.
6. Infiltrate *Agrobacterium* suspension into *N. benthamiana* leaves using a 1-mL needleless syringe.
7. Incubate *N. benthamiana* for 1 day under normal growth conditions (*see* **Note 6**).

3.2 Complementation Assay Using Codon-Shuffled NLR Gene

1. Spread *Agrobacterium* strain carrying binary vector for NLR protein expression on LB agar plates containing 50 μg/mL kanamycin and 50 μg/mL rifampicin. Incubate at 28 °C for 1–2 days.
2. **Steps 2–4** in Subheading 3.1.
3. Measure OD_{600} and adjust the OD_{600} with agroinfiltration buffer (*see* Tables 1, 2, and **Note** 7).
4. Infiltrate the *Agrobacterium* suspension into *N. benthamiana* leaves 1 day after RNAi experiment using a 1-mL needleless syringe.
5. Incubate *N. benthamiana* for 2 days under the normal growth conditions (*see* **Note 8**).

In a representative complementation assay, we overexpressed wild-type NbZAR1 fused with a 6xHA tag (NbZAR1^{WT}-6xHA), a codon-shuffled version (NbZAR1^{syn}-6xHA), and MADA motif mutants (NbZAR1^{L17E}-6xHA and NbZAR1$^{syn\text{-}L17E}$-6xHA) in *N. benthamiana* leaves by agroinfiltration 1 day after hpRNA: GUS or hpRNA:NbZAR1 expression (Fig. 1). Expression of hpRNA:NbZAR1 suppressed protein accumulation of NbZAR1^{WT}-6xHA and NbZAR1^{L17E}-6xHA, but not NbZAR1^{syn}-6xHA and NbZAR1$^{syn\text{-}L17E}$-6xHA (Fig. 1; also described in [7]). This suggests that epitope tag-fused proteins can be overexpressed in genetic complementation assays using *N. benthamiana* with suppressed endogenous gene expression.

3.3 Protein Extraction and Sample Preparation

1. Harvest 6 leaf disks from *N. benthamiana* and place them in a 2 mL microcentrifuge tube with two zirconia beads. Immediately freeze the samples in liquid nitrogen. You can store frozen samples at −80 °C.
2. Grind the frozen samples using ShakeMaster® AUTO at 1100 rpm for 1 min.
3. Add 200 μL of ice-cold protein extraction buffer to each sample tube on ice. Keep samples frozen until adding the extraction buffer to avoid protein degradation.

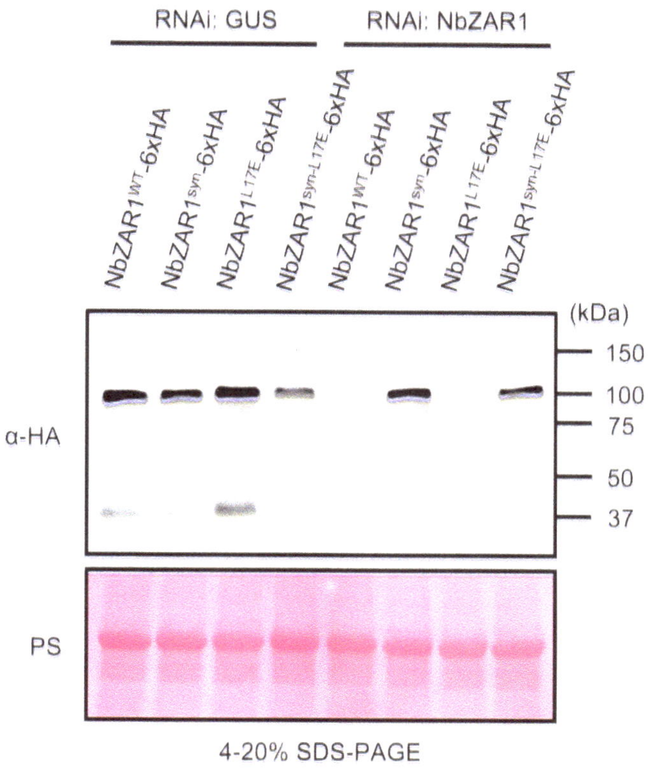

Fig. 1 Complementation assay using harpin RNA and codon-shuffled version of NbZAR1 (NbZAR1^{syn}) in *Nicotiana benthamiana*. NbZAR1^{WT}-6xHA, NbZAR1^{syn}-6xHA, NbZAR1^{L17E}-6xHA, and NbZAR1$^{syn\text{-}L17E}$-6xHA were overexpressed in *N. benthamiana* leaves by agroinfiltration 1 day after hpRNA:GUS (control) or hpRNA:NbZAR1 expression. Total proteins were extracted 2 days after second agroinfiltration and used for the western blot analysis. Loading control is indicated by Ponceau-S staining (PS)

4. Gently mix and incubate the samples on ice. Mix each tube every 2–3 min for 10 min (*see* **Note 9**).
5. Centrifuge at 20,000× *g* at 4 °C for 15 min.
6. Transfer 50 μL of the supernatant into new ice-cold 1.5 mL microcentrifuge tubes for BN-PAGE and SDS-PAGE, respectively (*see* **Note 10**).
7. For BN-PAGE, add 2.5 μL NativePAGE™ 5% G-250 Sample Additive, 25 μL NativePAGE™ 4× sample buffer, and 22.5 μL pure water. Prepare a premix of these reagents before adding. Keep samples on ice.
8. For SDS-PAGE, add 50 μL of 2× SDS-PAGE loading dye and heat at 70 °C for 10 min.

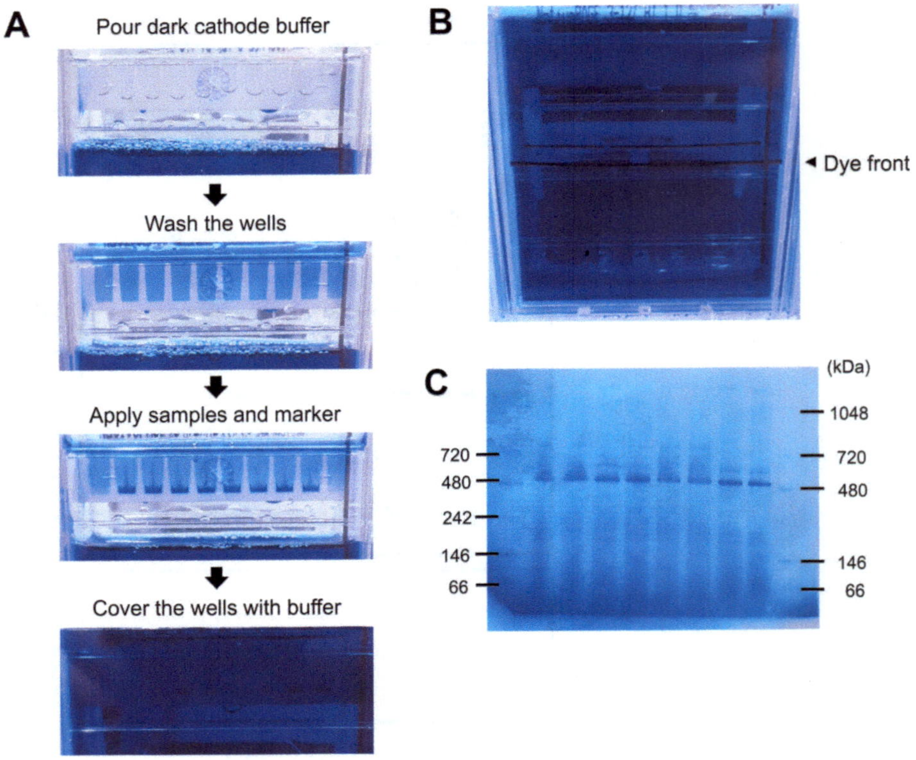

Fig. 2 Key steps in the BN-PAGE procedure. (**a**) Pictures showing each set-up process for BN-PAGE as described in Subheading 3.4, **steps 3–6**. (**b**) A picture showing the dye front reaching approximately one-third of the gel in Subheading 3.4, **step 7**. (**c**) A picture showing the markers visualized in Subheading 3.6, **step 7**

3.4 BN-PAGE

1. Rinse the NativePAGE™ 3–12% Bis-Tris Mini Protein Gel with pure water and remove the comb and bottom tape.
2. Set the gel in the XCell *SureLock*™ Mini-Cell tank.
3. Pour dark cathode buffer into the inner chamber until just below the wells (*see* Fig. 2a). Make sure no leakage to the outer chamber.
4. Wash the wells with dark cathode buffer by pipetting to replace the buffer (*see* Fig. 2a).
5. Load 5 μL of each sample and 2.5 μL NativeMark™ Unstained Protein Standard into the wells (*see* Fig. 2a, **Notes 11** and **12**).
6. Gently add darker cathode buffer into the inner chamber to fully cover the wells. Pour anode buffer into the outer chamber (*see* Fig. 2a and **Note 13**).
7. Run at 150 V for 30–40 min at 4 °C until the dye front reaches approximately one-third of the gel (*see* Fig. 2b).
8. Remove the dark cathode buffer from the inner chamber using a pump or pipette.
9. Add light cathode buffer to the inner chamber.
10. Restart the run at 200 V until the dye front reaches to the bottom of the gel.

3.5 SDS-PAGE

1. Rinse the Mini-PROTEAN® TGX™ Precast Gel with pure water and remove the comb.
2. Set the gel in the Mini-PROTEAN® Tetra Vertical Electrophoresis Cell tank.
3. Pour SDS-PAGE running buffer into the inner chamber, ensuring the wells are covered. Also, pour the running buffer into the outer chamber.
4. Wash the wells with running buffer by pipetting.
5. Load 15 μL of each sample and 5 μL of Precision Plus Protein Dual Color Standards.
6. Run at 100 V until the dye front reaches the bottom of the gel.

3.6 Western Blot of BN-PAGE

1. Soak the PVDF membrane in 100% methanol, then in transfer buffer. Soak two sets of filter paper in transfer buffer.
2. Cut the bottom edge of the gel and wells, and soak the gel in transfer buffer. Stack in the following order (bottom to top): filter paper, membrane, gel, filter paper (*see* **Note 14**). Remove any excess transfer buffer.
3. Perform protein transfer at 25 V for 30 min using the Trans-Blot® Turbo™ Transfer System.
4. Fix the proteins by incubating the membrane in 8% acetic acid for 15 min on a rocking platform.
5. Discard the acetic acid and wash the membrane four times with pure water for 30 s each.
6. Dry the membrane on paper for at least 40 min until fully dry.
7. To visualize markers, soak the membrane in 100% ethanol. Once the markers become visible, lift the membrane out of the ethanol and quickly mark positions with a pen (*see* Fig. 2c and **Note 15**). Re-soak in ethanol to remove background dye.
8. (Optional) Incubate the membrane in SDS solution for 15 min to expose hidden epitopes.
9. Wash the membrane twice with TBS-T for 5 min each.
10. Soak the membrane in blocking solution for 1 h.
11. Add anti-HA-Peroxidase antibody to the blocking solution (1: 5000 dilution), and incubate overnight at 4 °C or for 1 h at room temperature.
12. Wash the membrane twice with TBS-T for 30 s each.
13. Wash the membrane three times with TBS-T for 10 min each.
14. Apply SuperSignal™ West Dura Extended Duration Substrate evenly over the membrane and sandwich the membrane between the imaging films.

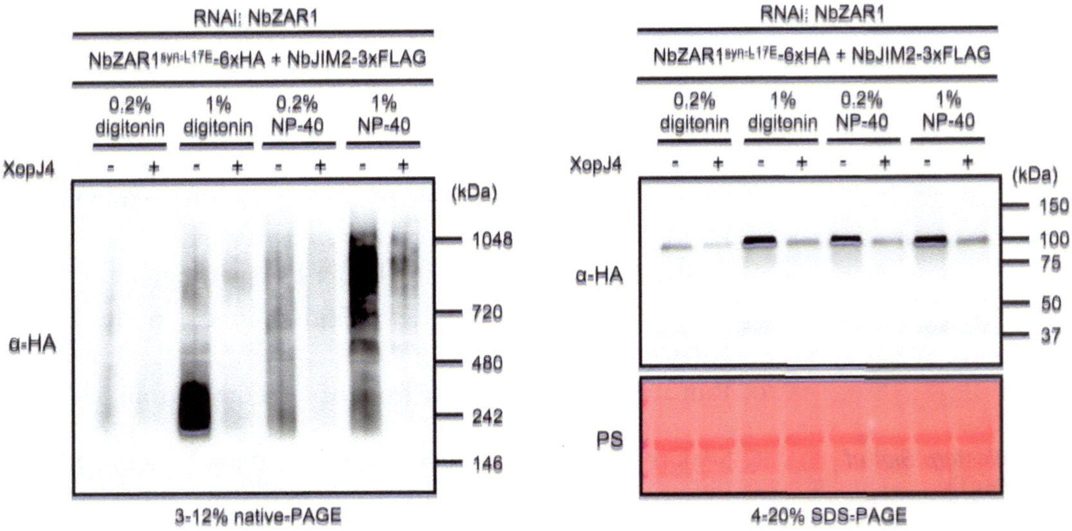

Fig. 3 BN-PAGE analysis of inactive and activated NbZAR1 proteins extracted under different detergent conditions. NbZAR1$^{syn\text{-}L17E}$-6xHA and NbJIM2-3xFLAG were co-expressed in *N. benthamiana* leaves by agroinfiltration with or without XopJ4, 1 day after hpRNA:NbZAR1 expression. 2 days after second agroinfiltration, total proteins were extracted using buffers containing different detergents. Samples were analyzed in parallel by BN-PAGE (left) and SDS-PAGE (right). Loading control is indicated by Ponceau-S staining (PS)

15. Detect chemiluminescence using LuminoGraph II EM (*see* **Note 16**).

In a representative BN-PAGE experiment, we co-overexpressed NbZAR1$^{syn\text{-}L17E}$-6xHA, NbJIM2-3xFLAG with or without XopJ4 in *N. benthamiana* leaves 1 day after hpRNA:NbZAR1 expression. 2 days later, total proteins were extracted using extraction buffers containing 0.2% or 1% digitonin, or 0.2% or 1% NP-40. BN-PAGE analysis revealed that band patterns of inactive and activated NbZAR1$^{syn\text{-}L17E}$-6xHA varied depending on detergent type and concentration (Fig. 3). Notably, extraction with 1% digitonin provided optimal conditions for analyzing NbZAR1 resistosome formation (Fig. 3). This highlights the critical importance of optimizing detergent conditions for successful BN-PAGE analysis of plant NLR proteins.

3.7 Western Blot of SDS-PAGE

1. Follow Subheading 3.6, **steps 1** and **2** using SDS-PAGE transfer buffer.
2. Perform protein transfer at 25 V for 7 min using the Trans-Blot® Turbo™ Transfer System.
3. Follow steps of Subheading 3.6, **steps 10–15**.
4. Soak the membrane in Ponceau-S staining solution until protein bands become visible.

5. Discard the Ponceau-S staining solution and wash the membrane 2–3 times with TBS-T for 30 s each.
6. Take a picture of the stained membrane.

4 Notes

1. All plasmids used in this protocol were described in a previous report [7]. We used a MADA motif mutant of NbZAR1 (NbZAR1^{L17E}) to suppress the cell death response triggered by activated NbZAR1 protein [7, 23]. This mutation strategy allows for the assessment of resistosome formation by activated plant CC-NLRs without cell death, which often interferes with biochemical experiments using plant cells [15, 23].
2. For biochemical experiments, we recommend using younger *N. benthamiana* leaves for agroinfiltration, as protein expression is typically more efficient in four- to five-week-old plants compared to those older than 6 weeks.
3. To minimize variability in protein expression caused by leaf-to-leaf differences during agroinfiltration, we recommend infiltrating two or more leaves from different plants and harvesting leaf disks from each infiltrated leaf.
4. The type and concentration of detergent in the extraction buffer are critical factors for preserving and detecting protein complexes. To optimize detergent conditions, we recommend testing various detergents and concentrations. Commonly used detergents include digitonin, NP-40, Triton X-100, and DDM.
5. The extraction buffer, anode buffer, cathode buffer, and transfer buffer should be used in a well-chilled state to maintain protein stability during the experiment. We recommend preparing these buffers 1 day in advance to allow for sufficient chilling and to save time on the day of the experiment.
6. In complementation assays, *Agrobacterium* strains carrying hpRNA constructs should be infiltrated into *N. benthamiana* leaves 1 day prior to overexpressing the proteins of interest. Simultaneous agroinfiltration of hpRNA and overexpression constructs often results in insufficient silencing of the endogenous gene.
7. Agroinfiltration enables the simultaneous expression of multiple proteins. To co-express several genes, mix *Agrobacterium* suspensions carrying each construct and adjust the final OD_{600} of each strain according to experimental requirements. We normally use an OD_{600} range of 0.1–0.5 per strain.

8. The timing and efficiency of protein expression by agroinfiltration can vary depending on the type of overexpression vector, the properties of each protein, and the plant growth conditions. Therefore, the optimal timing for sample harvesting should be determined individually for each target protein under your experimental conditions. In our experiments, samples are typically harvested at 2–3 days after agroinfiltration.
9. Ensure that samples are completely resuspended in the extraction buffer. Avoid using a vortex, as vigorous mixing may disrupt protein complexes.
10. BN-PAGE should be performed immediately after protein extraction to prevent dissociation of protein complexes. Once BN-PAGE has started, prepare SDS-PAGE. Throughout the process, keep all samples at 4 °C or on ice to preserve protein stability.
11. The sample loading volume should be kept small, as excessive volume may cause smeared bands and reduced resolution during BN-PAGE.
12. During marker visualization, the 1048 kDa marker can sometimes appear faint or difficult to detect. We recommend loading the marker into two separate wells.
13. If the volume of cathode or running buffer is insufficient, BN-PAGE may not run properly. Make sure that both the inner and outer chambers of the tank are adequately filled with their respective buffers, and verify that the electrophoresis is proceeding correctly.
14. During the transfer step, handle the bottom portion of the gel gently, as it is prone to breaking. Pouring a small amount of transfer buffer between each layer can help minimize bubble formation.
15. Markers, particularly those of high molecular weight, should be marked immediately, as they may only be visible for a few seconds.
16. Insufficient washing with TBS-T may lead to background signal during detection. In such cases, we recommend repeating from the washing step (Subheading 3.6, **step 13**) after detection to reduce background noise.

Acknowledgments

We are thankful to Sophien Kamoun (The Sainsbury Laboratory), Mauricio P. Contreras (The Sainsbury Laboratory), Lida Derevnina (University of Cambridge), and Yoshinobu Kato (University of Tokyo) for technical advice and valuable suggestions. We also thank our colleagues, especially Ryohei Terauchi's group members

(Kyoto University), for valuable discussions. HA is funded by the Japan Science and Technology Agency (JST), Precursory Research for Embryonic Science and Technology (JPMJPR21D1), and Japan Society for the Promotion of Science (JSPS) (23K20042, 24H00010, 25K02012). KH is funded by JSPS (25KJ1495).

References

1. Eubel H, Braun HP, Millar AH (2005) Blue-native PAGE in plants: a tool in analysis of protein-protein interactions. Plant Methods 1(1):11
2. Wittig I, Braun HP, Schägger H (2006) Blue native PAGE. Nat Protoc 1(1):418–428
3. Fiala GJ, Schamel WW, Blumenthal B (2011) Blue native polyacrylamide gel electrophoresis (BN-PAGE) for analysis of multiprotein complexes from cellular lysates. J Vis Exp 48:2164
4. Goodin MM, Zaitlin D, Naidu RA et al (2008) *Nicotiana benthamiana*: its history and future as a model for plant-pathogen interactions. Mol Plant-Microbe Interact 21(8):1015–1026
5. Bally J, Jung H, Mortimer C et al (2018) The rise and rise of *Nicotiana benthamiana*: a plant for all reasons. Annu Rev Phytopathol 56:405–426
6. Lawson AW, Macha A, Neumann U et al (2024) A versatile protocol for purifying recombinant proteins from *Nicotiana benthamiana* for structural studies. bioRxiv preprint. https://doi.org/10.1101/2024.10.13.618061
7. Harant A, Pai H, Sakai T et al (2022) A vector system for fast-forward studies of the HOPZ-ACTIVATED RESISTANCE1 (ZAR1) resistosome in the model plant *Nicotiana benthamiana*. Plant Physiol 188(1):70–80
8. Jones JD, Vance RE, Dangl JL (2016) Intracellular innate immune surveillance devices in plants and animals. Science 354(6316):aaf6395
9. Wang J, Hu M, Wang J et al (2019) Reconstitution and structure of a plant NLR resistosome conferring immunity. Science 364(6435):eaav5870
10. Hu M, Qi J, Bi G et al (2020) Bacterial effectors induce oligomerization of immune receptor ZAR1 *in vivo*. Mol Plant 13(5):793–801
11. Bi G, Su M, Li N et al (2021) The ZAR1 resistosome is a calcium-permeable channel triggering plant immune signaling. Cell 184(13):3528–3541.e12
12. Förderer A, Li E, Lawson AW et al (2022) A wheat resistosome defines common principles of immune receptor channels. Nature 610(7932):532–539
13. Zhao YB, Liu MX, Chen TT et al (2022) Pathogen effector AvrSr35 triggers Sr35 resistosome assembly via a direct recognition mechanism. Sci Adv 8(36):eabq5108
14. Feehan JM, Wang J, Sun X et al (2023) Oligomerization of a plant helper NLR requires cell-surface and intracellular immune receptor activation. Proc Natl Acad Sci U S A 120(11):e2210406120
15. Contreras MP, Pai H, Tumtas Y et al (2023) Sensor NLR immune proteins activate oligomerization of their NRC helpers in response to plant pathogens. EMBO J 42(5):e111519
16. Liu F, Yang Z, Wang C et al (2024) Activation of the helper NRC4 immune receptor forms a hexameric resistosome. Cell 187(18):4877–4889.e15
17. Madhuprakash J, Toghani A, Contreras MP et al (2024) A disease resistance protein triggers oligomerization of its NLR helper into a hexameric resistosome to mediate innate immunity. Sci Adv 10(45):eadr2594
18. Sakai T, Contreras MP, Martinez-Anaya C et al (2024) The *NRC0* gene cluster of sensor and helper NLR immune receptors is functionally conserved across asterid plants. Plant Cell 36(9):3344–3361
19. Yu H, Xu W, Chen S et al (2024) Activation of a helper NLR by plant and bacterial TIR immune signaling. Science 386(6728):1413–1420
20. Gong Z, Qi J, Hu M et al (2022) The origin and evolution of a plant resistosome. Plant Cell 34(5):1600–1620
21. Adachi H, Sakai T, Kourelis J et al (2023) Jurassic NLR: conserved and dynamic evolutionary features of the atypically ancient immune receptor ZAR1. Plant Cell 35(10):3662–3685
22. Schultink A, Qi T, Bally J et al (2019) Using forward genetics in *Nicotiana benthamiana* to uncover the immune signaling pathway mediating recognition of the *Xanthomonas perforans* effector XopJ4. New Phytol 221(2):1001–1009
23. Adachi H, Contreras MP, Harant A et al (2019) An N-terminal motif in NLR immune receptors is functionally conserved across distantly related plant species. elife 8:e49956

Chapter 10

Quantification of NLR-Induced Cell Death Using Rice Protoplasts

Ayaka Yoshihisa and Tsutomu Kawasaki

Abstract

Plant nucleotide-binding leucine-rich repeat receptors (NLRs) induce cell death by recognizing effectors secreted by pathogens into host cells. NLR activity can be evaluated by measuring the level of cell death. In *Nicotiana benthamiana*, this is typically done by co-expression of an NLR and its corresponding effector in the leaves via agroinfiltration. However, agroinfiltration is not suitable for protein expression in rice leaves. Here, we describe a method for preparing protoplasts from rice leaves and conducting a cell death assay using polyethylene glycol (PEG)-mediated transfection. This assay can also be applied to evaluate the activity of other cell death-inducing proteins.

Key words Rice, Protoplast, NLR, Effector, Cell death

1 Introduction

As an early response to pathogen attack, plants utilize plasma membrane-localized pattern recognition receptors (PRRs) to recognize conserved molecular patterns known as microbe or pathogen-associated molecular patterns (MAMPs/PAMPs) [1]. Recognition of these patterns triggers multiple immune responses, including mitogen-activated protein kinase (MAPK) cascade activation, reactive oxygen species (ROS) production, Ca^{2+} influx into the cytoplasm, lignin biosynthesis, and reinforcement of the cell wall. This initial immune response is referred to as pattern-triggered immunity (PTI) [2]. To overcome PTI, pathogens have evolved effector proteins that are secreted into host cells via pathogen-specific secretion systems. These effectors manipulate host cellular processes to create a favorable environment for pathogen proliferation. In response, plants have evolved intracellular immune receptors known as nucleotide-binding leucine-rich repeat receptor (NLR), which contain a central NB-ARC domain and a C-terminal LRR domain. Direct or indirect recognition of effectors

Tsutomu Kawasaki and Hirofumi Yoshioka (eds.), *Plant Immunity: Methods and Protocols*, Methods in Molecular Biology, vol. 3012, https://doi.org/10.1007/978-1-0716-5138-4_10,

by NLRs triggers a robust secondary immune response called effector-triggered immunity (ETI) [2, 3], which is often associated with a hypersensitive response (HR) characterized by localized cell death [3]. However, the molecular mechanisms underlying NLR-mediated immune activation remain unknown.

Transient expression in *N. benthamiana* leaves via agroinfiltration is a widely used method to analyze NLR-induced cell death. Indeed, many studies have utilized the *N. benthamiana* expression system to investigate the cell death-inducing activity of rice NLRs [4–6]. However, some rice NLRs do not trigger cell death in *N. benthamiana*, which suggests that rice-specific components may be required for their proper function. In such cases, a transient expression system using rice protoplasts offers a valuable alternative for analyzing the activity of these rice NLRs.

Recent studies have shown that certain NLRs form multimeric protein complexes known as resistosomes upon effector recognition. Resistosomes function as Ca^{2+}-permeable channels that pierce the plasma membrane, triggering Ca^{2+} influx, which in turn leads to cell death and transcriptional reprogramming [7, 8]. Transient expression system in *N. benthamiana* has also been used for the biochemical analysis of resistosome formation. Similarly, resistosome formation in rice cells may be investigated using a transient expression system with rice protoplasts.

Rice (*Oryza sativa*) is one of the most important cereal crops worldwide. Bacterial leaf blight, caused by *Xanthomonas oryzae* pv. *oryzae* (*Xoo*), is a major rice disease affecting rice production. *Xoo* delivers transcription-activator-like (TAL) effectors into rice cells, where they function as transcription factors to regulate host gene expression. The rice *Xa1* gene encodes an NLR with an N-terminal coiled coil (CC)—BED domain [9]. Xa1 recognizes *Xoo*TAL effectors and induces cell death [10, 11]. However, co-expression of Xa1 and TAL effectors in *N. benthamiana* does not induce cell death, suggesting rice-specific factors may be required for Xa1 function. To overcome this limitation, we developed a method for rice protoplast preparation and a corresponding cell death assay. This system provides a useful platform for analyzing immune responses mediated by rice NLRs.

2 Materials

Unless otherwise specified, all solutions were prepared using deionized water (ADVANTEC) and standard molecular biology regents and equipment.

2.1 Preparation of Rice Protoplasts

1. Deep-bottomed petri dish.
2. Microscalpel.
3. Vacuum pump.
4. Vacuum chamber.
5. 28 °C incubator.
6. 50 mL centrifuge tube.
7. Bovine serum albumin (BSA).
8. 600 mM Mannitol.
9. Digestion buffer: 600 mM Mannitol, 20 mM KCl, 10 mM $CaCl_2$, 20 mM MES-KOH (pH 5.7).
10. Cellulase solution: 1% (w/v) Cellulase "ONOZUKA" RS (YAKULT), 0.5% (w/v) Macerozyme R-10 (YAKULT), 0.1% (w/v) BSA, 10 mM β-mercaptoethanol (Nacalai), which are dissolved in digestion buffer.
11. Swinging-bucket rotor centrifuge.
12. 70 μm cell strainer (product number 352350, Falcon®).
13. W5 buffer: 150 mM NaCl, 125 mM $CaCl_2$, 5 mM KCl, 2 mM MES (pH 5.7).
14. MMG buffer: 400 mM Mannitol, 15 mM $MgCl_2{\cdot}6H_2O$, 4 mM MES (pH 5.7).

2.2 Transformation

1. Swinging-bucket rotor centrifuge.
2. 28 °C incubator.
3. PEG solution: 40% (w/v) polyethylene glycol 4000 (Sigma), 200 mM Mannitol, 100 mM $CaCl_2$.
4. W5 buffer: (same as in Subheading 2.1).

2.3 Cell Death Assay

1. Swinging-bucket rotor centrifuge.
2. 28 °C incubator.
3. Luminometer (TriStar^2 LB942, Berthold Technologies).
4. 96-well half-area plate.
5. Nano-Glo@ Live cell Reagent (Promega).

3 Methods

3.1 Preparation of Rice Protoplasts

1. Use rice plants 10–14 days after sowing (*see* **Note 1**). Plant growth condition: 13-h light (28 °C)/11-h dark (24 °C) cycle under white fluorescent light in a growth chamber (*see* **Note 2**).
2. Harvest approximately 10 seedlings by cutting them at the base of the stem (*see* **Note 3**).

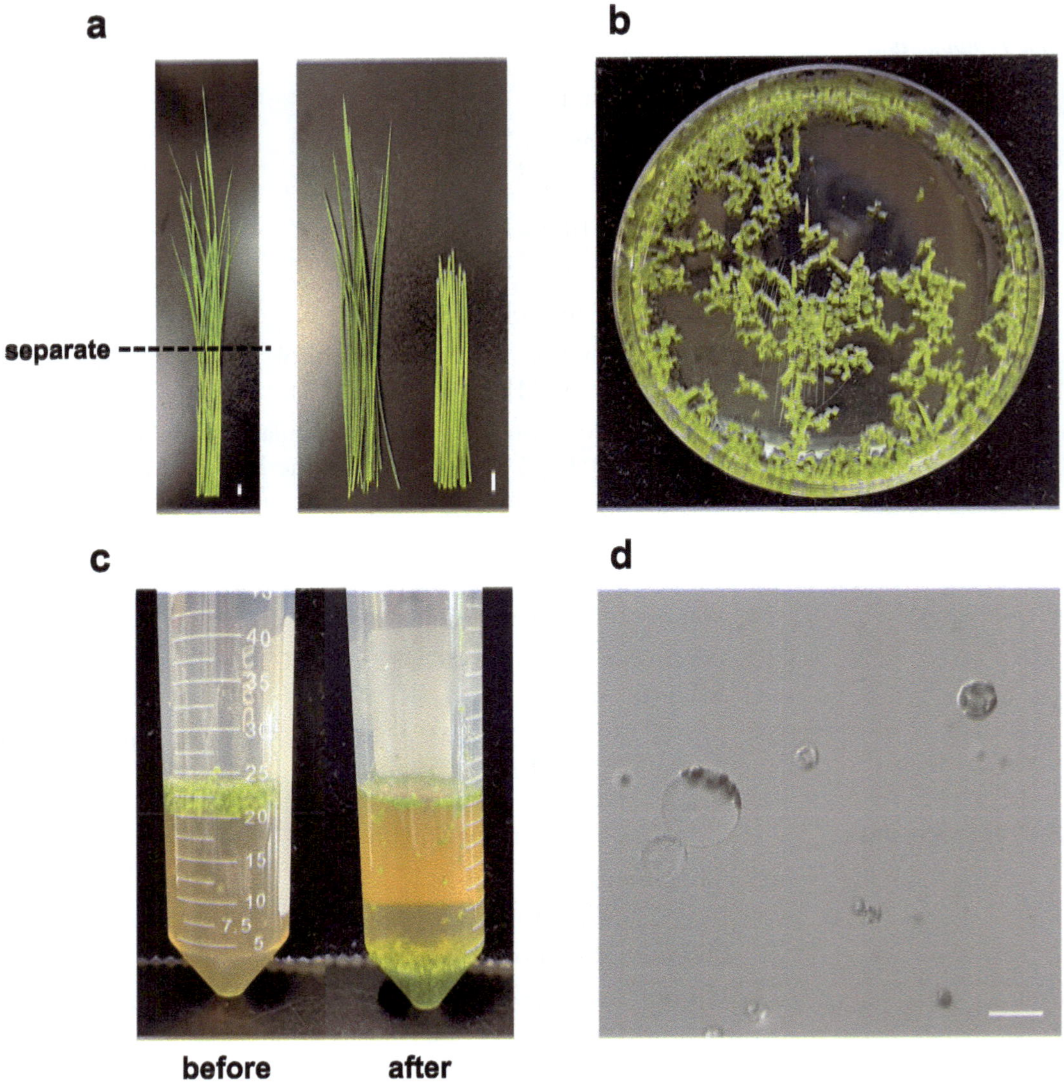

Fig. 1 Purification of rice protoplasts. (**a**) Separation of rice leaf blades from stems and sheaths. Scale bar: 1 cm. (**b**) Stems and sheaths chopped into 0.5–1 mm pieces. (**c**) The tissues before (left) and after (right) degassing using a vacuum pump. (**d**) Bright-field image of purified rice protoplasts. Scale bar: 20 μm

3. Remove the leaves and retain only the stem and sheath portions by cutting approximately 10 cm above the base (*see* **Note 4** and Fig. 1).
4. Chop the stems and sheaths into 0.5–1 mm piece using a microscalpel while submerged in 20 mL of 600 mM mannitol in a deep-bottomed petri dish (*see* **Note 5**).
5. Incubate the chopped tissue in the mannitol solution for 10 min in the dark.

6. Carefully remove and discard the mannitol solution.
7. Add 20 mL of the cellulase solution to the petri dish containing the plant tissue.
8. Degas the mixture using a vacuum pump until air bubbles are released from the tissue.
9. Slowly return the mixture to atmospheric pressure (*see* **Note 6**).
10. Repeat **steps 8** and **9** (vacuum/degas cycle) approximately five times, until the tissue sinks to the bottom of the cellulase solution.
11. Incubate the mixture at 28 °C with gentle side-to-side shaking at 40 rpm for 3 h (*see* **Note 7**).
12. Set up a 70 μm cell strainer over a 50 mL Falcon tube and wet the strainer with 500 μL of W5 buffer.
13. After incubation, filter the protoplast mixture through the strainer.
14. Transfer any remaining plant tissue from the tube onto the strainer.
15. Wash the retained tissue on the strainer with 1 mL of W5 buffer (*see* **Note 8**).
16. Dilute the filtrate with W5 buffer to a final volume of 40 mL.
17. Centrifuge at 23 °C, 200× *g*, for 10 min (slow deceleration).
18. Discard the supernatant and gently resuspend the pellet in 20 mL of W5 buffer using a pipette with a bent tip.
19. Centrifuge at 23 °C, 200× *g*, for 5 min (slow deceleration).
20. Repeat **steps 18–19** once.
21. Gently resuspend the final pellet in 20 mL of MMG buffer using a pipette.
22. Centrifuge at 23 °C, 200× *g*, for 5 min (slow deceleration).
23. Discard the supernatant and resuspend in 5 mL of MMG buffer.
24. Count the number of protoplasts using a hemocytometer and adjust the concentration to 2.0–3.0 × 10^6 cells per 100 μL.

3.2 Transformation

1. Add up to 20 μg of plasmid DNA to a 2 mL microcentrifuge tube (total DNA volume ≤10 μL).
2. Add 100 μL of protoplast suspension and mix by vortexing (*see* **Note 9**).
3. Add 110 μL of PEG solution and mix thoroughly by vortexing until the solution becomes homogeneous.
4. Incubate the mixture at room temperature for 10 min.

5. Add 1 mL of W5 buffer to stop the reaction.
6. Centrifuge at 23 °C, 200× *g*, for 10 min (slow deceleration).
7. Discard the supernatant and add 1 mL of fresh W5 buffer.
8. Centrifuge again at 23 °C, 200× *g*, for 5 min.
9. Repeat **steps 7–8** two more times.
10. Finally, resuspend the pellet in 1 mL of W5 buffer.
11. Incubate at 28 °C in the dark with gentle side-to-side shaking (60 rpm) for 18 h (*see* **Note 10**).

3.3 Cell Death Assay

1. After 18 h incubation, centrifuge the protoplast suspension at 28 °C, 200× *g* for 10 min.
2. Remove the supernatant and resuspend the pellet in 25 μL of W5 buffer by vortexing.
3. Add 5 μL of Nano-Glo@ Live Cell Reagent to each well of a 96-well plate.
4. Transfer 20 μL of the resuspended protoplast suspension into each corresponding well.
5. Immediately measure luciferase activity using a luminometer (TriStar^2).

In a representative cell death assay (Fig. 2), we co-expressed AvrXa7, a TAL effector from *Xanthomonas oryzae* pv. *oryzae*, with NanoLuc luciferase in protoplasts derived from the rice cultivar Kogyoku, which carries the *Xa1* gene. Compared with the negative control (GUS), luciferase activity was reduced upon expression of AvrXa7, indicating that cell death was induced through recognition of AvrXa7 by Xa1.

4 Notes

1. Protoplasts can also be prepared from older rice plants (up to approximately 3 weeks after sowing). However, the yield and transformation efficiency may be reduced.
2. If rice seedlings are grown under suboptimal conditions, they tend to become elongated. Protoplasts prepared from such plants often exhibit low transformation efficiency.
3. As described in Subheading 3.2, **step 1**, each transformation uses 100 μL of protoplast suspension (2.0–3.0 × 10^6 cells). To prepare 15 mL of protoplast suspension (sufficient for 150 samples), 10 seedlings are used.

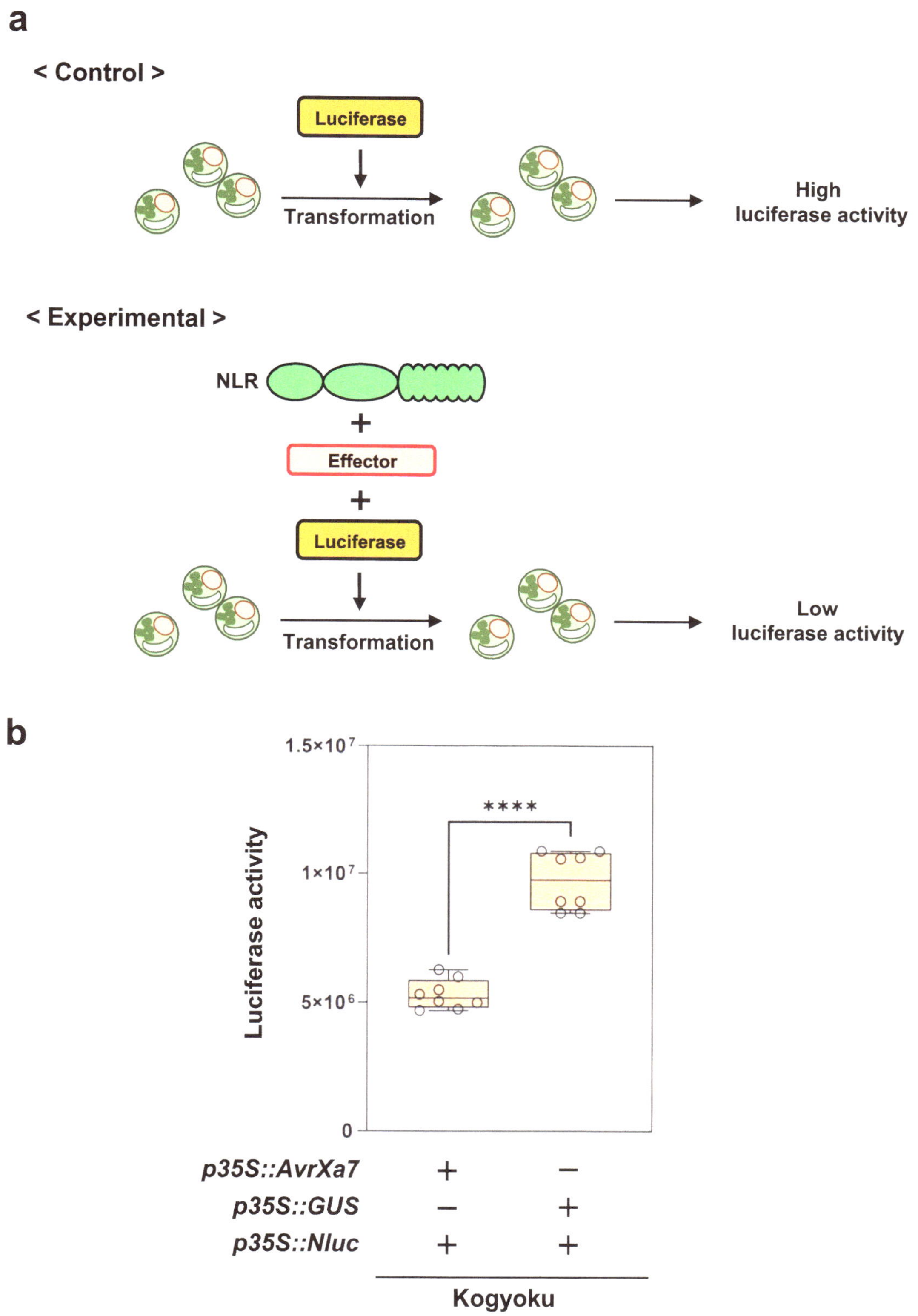

Fig. 2 Principle and example of cell death analysis in rice protoplast. (**a**) Schematic of the assay principle. Rice protoplasts are transformed via PEG-mediated transformation with constructs expressing luciferase either

4. Although stem and sheath are preferred, rice leaves can also be used. Note that protoplasts derived from leaves may be more fragile or prone to damage.
5. As described in Subheading 3.1, vacuum degassing allows the enzyme solution to infiltrate the rice tissues. However, if the tissue pieces are too large, the solution may not penetrate sufficiently. Conversely, if the tissues are cut into pieces smaller than 0.5 mm, the fragments may become crushed.
6. It is crucial to gradually restore atmospheric pressure after the degassing process.
7. Gentle shaking during enzymatic treatment improves the efficiency of the reaction.
8. Gently press the tissue with a spoon or similar tool to extract residual buffer thoroughly. Incomplete extraction may result in a lower protoplast yield.
9. Vortex the sample five times for 1 s each.
10. In general, protoplasts are incubated for 18 h following transformation; however, the optimal incubation period may vary depending on the protein of interest. Therefore, optimization of incubation conditions is recommended.

Acknowledgments

This research was supported by Grants-in-Aid for Scientific Research(A)(24H00515) to T.K.; by Grant-in-Aid for JSPS Fellows (24KJ2161) and a grant from the Yanmar Environmental Sustainability Support Association to A.Y.

Fig. 2 (continued) alone (control) or together with an NLR and its corresponding pathogen effector (experimental). In the control sample, high luciferase activity is observed due to the absence of cell death. In contrast, co-expression of the NLR and effector induces cell death, resulting in reduced luciferase activity. (**b**) Example: Xa1-mediated cell death triggered by recognition of the *Xanthomonas* TAL effector AvrXa7. Constructs expressing NanoLuc luciferase with or without AvrXa7 were introduced into protoplasts derived from the rice cultivar Kogyoku, which carries the *Xa1* gene. A GUS construct was used as a negative control. At 18 h post-transformation, luciferase activity was measured using a luminometer and the Nano-Glo Live Cell Assay System. The result indicates that recognition of AvrXa7 by Xa1 triggered cell death in rice protoplast, as reflected by a decrease in luciferase signal

References

1. Couto D, Zipfel C (2016) Regulation of pattern recognition receptor signalling in plants. Nat Rev Immunol 16:537–552. https://doi.org/10.1038/nri.2016.77
2. Dangl JL, Horvath DM, Staskawicz BJ (2013) Pivoting the plant immune system from dissection to deployment. Science 341:746–751. https://doi.org/10.1126/science.1236011
3. Contreras MP et al (2023) NLR receptors in plant immunity: making sense of the alphabet soup. EMBO Rep 24:e57495. https://doi.org/10.15252/embr.202357495
4. Liu F et al (2024) Activation of the helper NRC4 immune receptor forms a hexameric resistosome. Cell 187:4877–4889. e4815. https://doi.org/10.1016/j.cell.2024.07.013
5. Sakai T et al (2024) The NRC0 gene cluster of sensor and helper NLR immune receptors is functionally conserved across asterid plants. Plant Cell 36:3344–3361. https://doi.org/10.1093/plcell/koae154
6. Chen R et al (2025) A wheat tandem kinase activates an NLR to trigger immunity. Science 387:1402–1408. https://doi.org/10.1126/science.adp5034
7. Contreras MP et al (2023) Sensor NLR immune proteins activate oligomerization of their NRC helpers in response to plant pathogens. EMBO J 42:e111519. https://doi.org/10.15252/embj.2022111519
8. Jones JDG, Staskawicz BJ, Dangl JL (2024) The plant immune system: from discovery to deployment. Cell 187:2095–2116. https://doi.org/10.1016/j.cell.2024.03.045
9. Yoshimura S et al (1998) Expression of Xa1, a bacterial blight-resistance gene in rice, is induced by bacterial inoculation. Proc Natl Acad Sci USA 95:1663–1668. https://doi.org/10.1073/pnas.95.4.1663
10. Yoshihisa A et al (2022) The rice OsERF101 transcription factor regulates the NLR Xa1-mediated immunity induced by perception of TAL effectors. New Phytol 236:1441–1454. https://doi.org/10.1111/nph.18439
11. Yoshihisa A et al (2025) Essential role of rice ERF101 in the perception of TAL effectors and immune activation mediated by the CC-BED NLR Xa1. Plant Cell Rep 44:49. https://doi.org/10.1007/s00299-025-03436-7

References

[illegible]

Chapter 11

Split-TurboID-Based Proximity Labeling for Probing the Composition of Spatiotemporally Defined Protein Complexes In Planta

Zhiyan Wen, Xinyu Zhang, Yuexiao Li, Dingliang Zhang, Savithramma P. Dinesh-Kumar, and Yongliang Zhang

Abstract

Proteins frequently interact with other proteins to execute specific biological processes. Identifying protein-protein interactions (PPIs) is essential for understanding the molecular pathways underlying various physiological functions. Proximity labeling (PL), particularly the TurboID-based PL approach, has been widely utilized to identify PPIs across different organisms in recent years. For the first time, we applied the TurboID-based PL method to detect PPIs in plants. We recently demonstrated that the TurboID complementation approach, referred to as split-TurboID, can be used to screen for binding partners of defined protein complexes *in planta*. Here, we provide a detailed protocol for performing split-TurboID-based PL assays in *Nicotiana benthamiana* plants.

Key words Protein-protein interactions (PPIs), Proximity labeling, Split-TurboID, LC-MS/MS, SGT1-HSP90 protein complex

1 Introduction

Protein-protein interactions (PPIs) are fundamental to numerous biological processes, making the development of methods to identify them essential for understanding the mechanisms underlying various physiological processes. Recently, proximity labeling (PL) technology has emerged as a powerful tool for exploring weak, transient, or hydrophobic PPIs in living cells and mapping protein interaction networks. This method offers advantages over traditional approaches to studying PPIs, such as co-immunoprecipitation (co-IP) combined with mass spectrometry (MS) and yeast two-hybrid assays [1–3]. The widespread use of PL technology is driven by continuous optimization and advancements in PL enzymes. TurboID, derived from the directed evolution of the *Escherichia coli* biotin ligase (BirA), has gained traction in animals

Tsutomu Kawasaki and Hirofumi Yoshioka (eds.), *Plant Immunity: Methods and Protocols*, Methods in Molecular Biology, vol. 3012, https://doi.org/10.1007/978-1-0716-5138-4_11,

[2], plants [4, 5], and microorganisms [6] due to its small size and minimal impact on the functionality of the target protein.

To enhance the specificity and flexibility of this technique, split-TurboID technology was developed by dividing TurboID into two inactive fragments. This system is especially useful for identifying components of known protein complexes or proteins associated with membrane contact sites. By fusing two interacting bait proteins to the N-terminal and C-terminal fragments of TurboID, respectively, their interaction within cells brings the separated TurboID fragments close together, reconstituting a functional TurboID enzyme. This reconstituted enzyme biotinylates proteins near the protein complexes, enabling the identification of additional components of spatiotemporally defined protein complexes (Fig. 1a).

Although split-TurboID has been successfully utilized to study PPIs in mammalian cells [7, 8], its application in plant cells is still in the early stages. Recently, we developed a toolbox for TurboID- and split-TurboID-based PL assays to investigate plant PPIs [9]. Using the Suppressor of the G2 allele of SKP1 (SGT1) and Heat shock protein 90 (HSP90) complex as a model, we identified several novel components of the SGT1-HSP90 complex [10], further demonstrating the effectiveness of the split-TurboID system for exploring molecular interactions *in planta*.

The application of split-TurboID for analyzing protein complexes in *Nicotiana benthamiana* involves the following steps: bait proteins fused with either the N- or C-terminal fragments of TurboID are transiently co-expressed in *N. benthamiana* leaves. After cell lysis and desalting, the leaf extracts undergo streptavidin pull-down using beads. The enriched proteins are enzymatically digested, labeled with tandem mass tag (TMT) reagents, and identified through liquid chromatography with tandem mass spectrometry (LC-MS/MS). Bioinformatics analysis reveals potential new components of known protein complexes (Fig. 1b).

2 Materials

2.1 Plant Material

1. Four- to six-week-old *N. benthamiana* plants. Grow *N. benthamiana* seeds in a climate-controlled chamber at 23 °C under a 16 h/8 h, light/dark cycle.

2.2 Reagents

1. RIPA lysis buffer: 50 mM Tris–HCl (pH 7.5), 500 mM NaCl, 1 mM EDTA (pH 8.0), 1% IGEPAL®CA-630 (*v/v*), 0.1% SDS (*w/v*), 0.5% sodium deoxycholate (*w/v*). Store at 4 °C. Add 1 mM dithiothreitol (DTT) and two tablets of cOmplete™ Protease Inhibitor Cocktail (Sigma, Cat. No. 11697489001) per 100 mL before use.

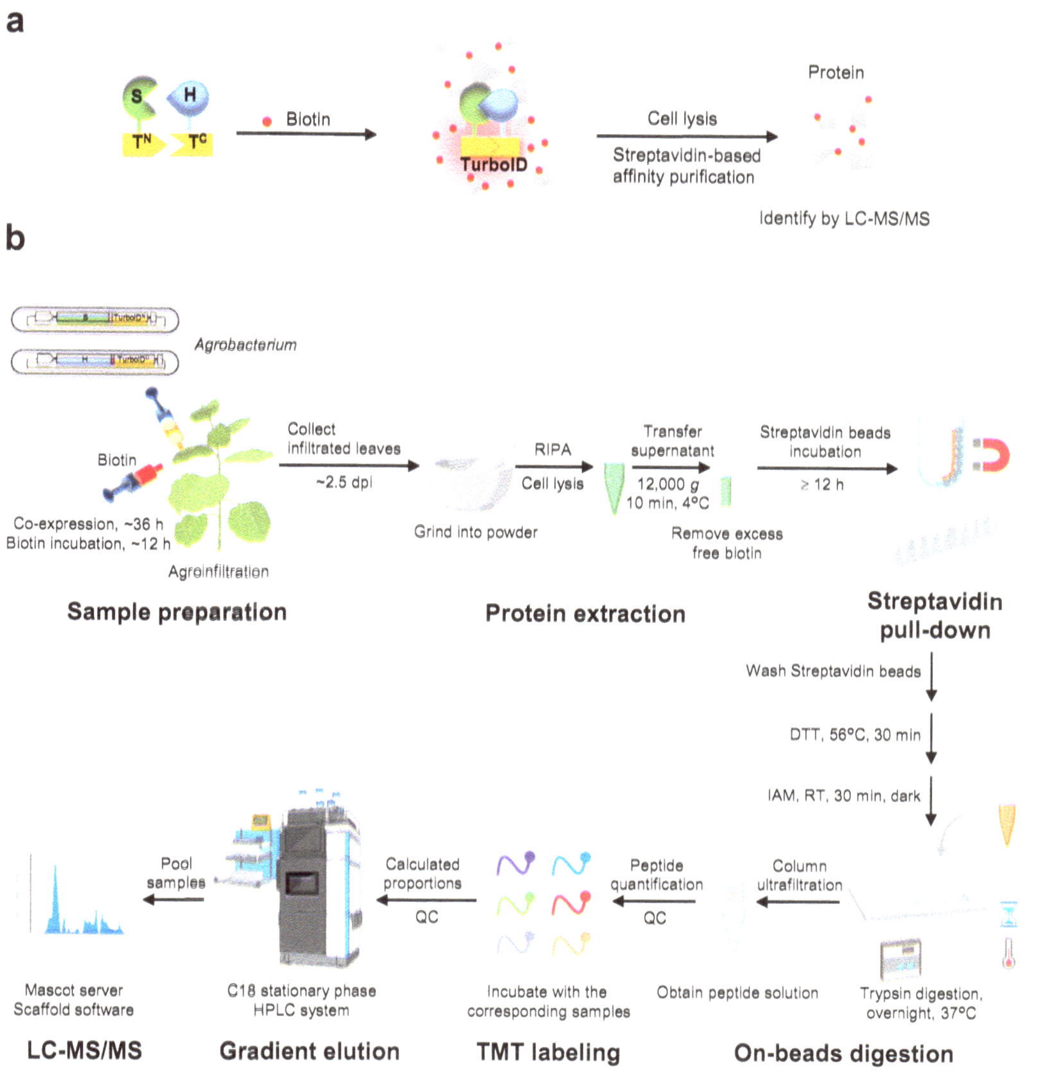

Fig. 1 Schematic of the split-TurboID-based PL assay for investigating additional components of known protein interactions. (**a**) Principle of the split-TurboID-based PL method. TN and TC stand for the N- and C-terminal fragments of TurboID. S and H represent a pair of known interacting proteins, SGT1 and HSP90. Red dots show biotin molecules, while gray shapes indicate biotinylated proteins. (**b**) Workflow of the split-TurboID-based PL assay. Agrobacterium cultures carrying split-TurboID constructs were infiltrated into *N. benthamiana* leaves. At 36 h post-infiltration, biotin was added to the leaves and incubated for 12 h at room temperature. Leaf samples were then harvested, ground into a fine powder, and lysed in RIPA buffer for 30–40 min. This was followed by centrifugation at 12,000 *g* for 10 min at 4 °C to obtain the supernatant. Excess biotin was removed using a desalting column. The eluate was incubated with streptavidin beads for 12 h. The beads were then sequentially washed with wash buffer, Tris–HCl, and ammonium bicarbonate. Proteins bound to streptavidin beads were reduced with DTT, alkylated with IAM, and digested overnight with trypsin. The resulting peptides were quantified, labeled with TMT, fractionated by gradient elution, and identified by LC–MS/MS

2. Wash buffer I: 2% SDS (*w/v*).
3. Wash buffer II: 50 mM HEPES (pH 7.5), 500 mM NaCl, 1 mM EDTA (pH 8.0), 0.1% deoxycholic acid (*w/v*), and 1% Triton X-100.
4. Wash buffer III: 10 mM Tris–HCl (pH 7.5), 250 mM LiCl, 1 mM EDTA (pH 8.0), 0.1% deoxycholic acid (*w/v*), and 1% IGEPAL®CA-630 (*v/v*) (*see* **Note 1**).
5. 5× Bradford regent (200 mL): 100 mg Coomassie Brilliant Blue Dye G-250, 47 mL methanol, 100 mL phosphoric acid (85 wt. % in H_2O), 53 mL Milli-Q H_2O. Keep away from light.
6. 50 mM ammonium bicarbonate.
7. Zeba Spin Desalting Column (Thermo Fisher Scientific, Cat. No. 89893).
8. Dynabeads™ MyOne™ Streptavidin C1 (Invitrogen, Cat. No. 65001).
9. Immobilon-P PVDF membrane (Millipore, Cat. No. IPVH00010).
10. 10,000 MWCO HY ultrafiltration products of VIVACON® 500 (Sartorius Stedim Biotech GmbH, Cat. No. VN01H02).
11. Pierce Quantitative Colorimetric Assay (Thermo Fisher Scientific, Cat. No. 23275).
12. TMTsixplex™ Label Reagent (Thermo Fisher Scientific, Cat. No. 90066).

3 Methods

3.1 Vector Construction

1. Clone a well-characterized pair of interacting proteins, SGT1 and HSP90, then construct entry vectors by inserting them into the pDONOR207 vector (Invitrogen) via Gateway BP recombination.
2. Transform SGT1 and HSP90 fragments into corresponding destination vectors, resulting in p35S:SGT1-TurboID^N and p35S:HSP90-TurboID^C [9], respectively, which can be used for transient overexpression in *N. benthamiana* by Gateway LR recombination.

3.2 Agroinfiltration

1. *N. benthamiana* leaves were infiltrated with *Agrobacterium* mixtures expressing SGT1-TurboID^N at a final OD_{600} of 0.5 and HSP90-TurboID^C at a final OD_{600} of 0.5, with Citrine-TurboID as the control group at a final OD_{600} of 1.0.
2. At 36 h post-infiltration, leaves were infiltrated with 200 μM biotin (in 10 mM $MgCl_2$) and incubated for ~12 h. Each treatment group included three biological replicates (*see* **Note 2**).

3.3 Protein Extraction

1. Collect the infiltrated leaf samples and flash-freeze in liquid nitrogen. Store the frozen leaf samples at −80 °C until use.
2. Grind the leaves into powder, then weigh 0.35 g of the powder into a 2 mL tube; prepare two tubes with a total of ~0.7 g.
3. Add 700 μL of RIPA lysis buffer to each tube.
4. Vortex for 10 min to ensure complete dissolution of the powder in the buffer.
5. Keep the tubes incubated on ice for 30–40 min, and invert them multiple times during this period. Then centrifuge at 12,000 *g* for 10 min at 4 °C (*see* **Note 3**).
6. Centrifuge Zeba Spin Desalting Columns (one sample per column) at 1000 *g* for 2 min at 4 °C to remove the protective solution from the desalting columns.
7. Equilibrate the desalting column by adding 5 mL RIPA extraction buffer, then centrifuge at 1000 *g* for 2 min at 4 °C (*see* **Note 4**).
8. Repeat **step** 7 for additional twice.
9. Place the columns into new 50 mL centrifuge tubes, and slowly add the supernatant derived from **step 5** into the desalting columns using a pipette. Allow the solution to flow naturally until the liquid ultimately enters the column (*see* **Note 5**).
10. Add 200 μL of RIPA extraction buffer to the column.
11. Centrifuge at 1000 *g* for 2 min at 4 °C to collect protein extract. Aliquot 40 μL of the extract as 'Input' for subsequent immunoblot analysis (Fig. 2a) (*see* **Note 6**).

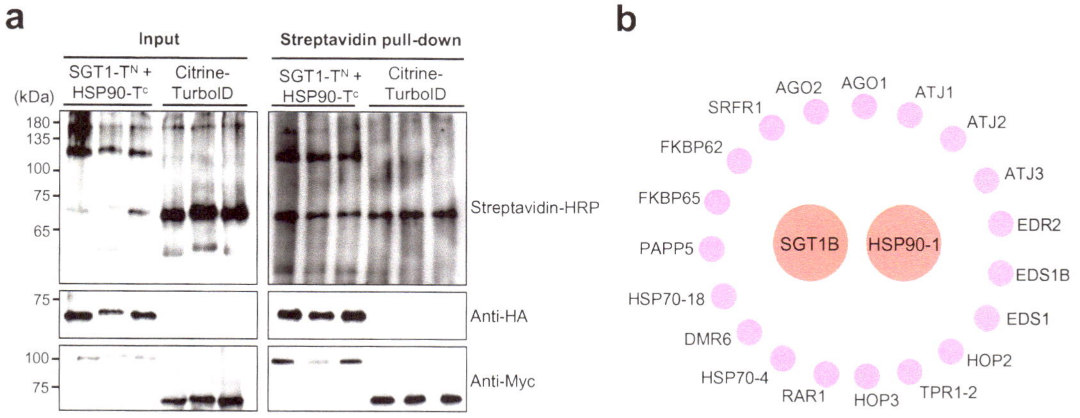

Fig. 2 Results of the split-TurboID-based PL assay. (**a**) Immunoblot analysis confirming the expression and biotinylation of split-TurboID fusion proteins in input and pull-down samples. Citrine-TurboID served as a control. Biotinylated proteins were detected with streptavidin-HRP, while SGT1-T^N and HSP90-T^C were identified using anti-HA and anti-Myc antibodies, respectively. (**b**) Interaction network identified by the split-TurboID-based PL assay. Nodes represent individual proteins labeled with their names, and edges indicate interactions predicted by STRING and visualized in Cytoscape

3.4 Quantification of the Desalted Protein Extracts

Quantify each sample's protein amounts via the Bradford assay [11] and measure absorbance at 595 nm (UV-Vis range) using an ELISA plate reader.

1. Prepare six Bovine Serum Albumin (BSA) reference samples (1 mg/mL, 0.8 mg/mL, 0.6 mg/mL, 0.4 mg/mL, 0.2 mg/mL, Blank in 50 μL total volume) for the standard curve.
2. Pipet 5 μL samples per 50 μL total volume.
3. Add 2.5 mL of 1× Bradford reagent to each tube and gently vortex-mix.
4. Measure the A_{595} of the samples and the references against the blank after 10 min of standing still. Pipette each sample into the 96-well plate in triplicate to minimize error through averaging.
5. Based on the measurement results, transfer the extracts containing equal protein amounts into the corresponding 2 mL tubes. This typically normalizes the total protein by using the sample with the lowest concentration as a reference and adjusting it to a final volume of 2 mL.

3.5 Streptavidin Pull-Down

1. Pipet streptavidin beads (100 μL per sample) into a 2 mL tube containing 1 mL RIPA extraction buffer (*see* **Note 7**).
2. Adsorb the beads for 2 min with a magnetic rack and discard the solution.
3. Equilibrate the beads with 1 mL pre-chilled RIPA extraction buffer (*see* **Note 8**).
4. Repeat **steps 2** and **3** three times.
5. Adsorb the beads on a magnetic rack for 2 min. Mix the desalted protein extract derived from **step 5** of Subheading 3.4 with the beads.
6. Incubate overnight (~12 h) at 4 °C on a rolling shaker.
7. Place the beads on the magnetic rack for 5 min at room temperature (RT), then remove the supernatant using a pipette.
8. Add 1.7 mL of wash buffer I to resuspend the beads.
9. Rotate for 8 min at RT on the rolling shaker. Repeat **step 7**.
10. Add 1.7 mL of wash buffer II to resuspend the beads. Repeat **steps 9** and 7.
11. Add 1.7 mL of wash buffer III to resuspend the beads. Repeat **steps 9** and 7.
12. To resuspend the beads, add 1.7 mL of 50 mM Tris–HCl (pH 7.5). Repeat **steps 9** and 7.
13. Transfer the beads with 600 μL of 50 mM Tris–HCl (pH 7.5) to a new 1.5 mL tube.

14. Add 600 μL of 50 mM Tris–HCl (pH 7.5) to the original 2 mL tube to transfer the residual beads to ensure maximum recovery. Repeat **steps 9** and 7.
15. Resuspend the beads by adding 1 mL of 50 mM ammonium bicarbonate buffer and repeat **steps 9** and 7.
16. Repeat **step 15** more than six times to wash the beads. Perform an immunoblot assay after taking out 50 μL of resuspended beads as the Streptavidin pull-down (Fig. 2a), and the resting beads were used for subsequent LC-MS/MS analysis.

3.6 On-Beads Digestion

1. Adsorb the Streptavidin beads for 5 min with a magnetic rack and then remove the supernatant using a pipette.
2. Wash beads with 200 μL of 50 mM triethylammonium bicarbonate (TEAB) on the constant temperature oscillating metal bath for ~2 min at RT. Repeat **step 1**.
3. Add 100 μL of 5 mM DTT in 50 mM TEAB per tube, then react at 56 °C for 30 min. Repeat **steps 2** and **1**.
4. Add 100 μL of 20 mM IAM in 50 mM TEAB per tube, then incubate at RT for 30 min under dark conditions. Repeat **steps 2** and **1**.
5. Add 1 μg (100 μL) of trypsin dissolved in 50 mM TEAB (pH 8.0–8.5) to each tube, followed by digestion overnight at 37 °C.
6. Add 50 μL of 50 mM TEAB to the 10,000 MWCO HY ultrafiltration products of VIVACON® 500 and centrifuge at maximum speed (16,160 *g*) for 10–20 min until no liquid remains visible on the filter.
7. After the magnetic rack enriches the Streptavidin beads, pipette the supernatant into ultrafiltration tubes and centrifuge at maximum speed for at least 30 min to collect the peptide enzymatic solution.
8. Each 5 μL sample is mixed with 15 μL of Milli-Q H_2O for peptide quantification.
9. Peptide quantification was performed using the Pierce Quantitative Colorimetric Assay following the protocol provided in the kit.
   (a) Prepare Standards
   BSA: 500 ng, 250 ng, 125 ng, 62.5 ng, 31.25 ng; blank.
   (b) Prepare Working Reagent (WR)
   Use the following formula to determine the total volume of WR required:
   [(# of standards) + (# of unknowns)] × (# of replicates) × (0.18 mL of WR per sample) = total volume WR required. #: number; ×: multiply.

Colorimetric Peptide Assay Reagent A: B: C = 50: 48: 2

After preparation, the WR can be kept at RT for ≤30 min.

(c) Procedure

Each 20 μL sample replicate or standard was mixed with 180 μL of WR at 37 °C for 15 min. Then, 200 μL of each mixture was pipetted into a well of a 96-well microplate. Finally, the absorbance value of each mix was obtained using a 96-well microplate reader compatible with the absorbance measurement at 480 nm (UV-Vis range).

10. Samples were taken for quality control of LC-MS/MS, and an injection content of ~600 ng was used to verify protein extraction quality and enzymatic hydrolysis efficiency.

3.7 TMT Labeling

Once the quality control results are verified, the peptide segments are TMT labeled, with TMTsixplex used here as an example.

1. 20 μg of each sample was supplemented with 50 mM TEAB to standardize the volume to 70 μL.
2. Add corresponding 40 μL of TMTsixplex^{TM} Label Reagent dissolved in acetonitrile (ACN) to the sample according to the experimental design, and the reaction mixture was incubated at RT for 1.5 h (*see* **Note 9**).
3. Add 20 μL of 10% formic acid (FA) to each tube and incubate at RT for 15 min to terminate the labeling reaction.
4. Take 5 μL from each sample and mix them in a new 1.5 mL tube. Dry the mixture by vacuum centrifugation at 45 °C.
5. Redissolve the mixture powder by adding 20 μL of 0.1% FA solution. Then, take the redissolved mixture to calculate the blending ratio of TMT labeling samples, and the injection content for LC-MS/MS is ~700 ng (*see* **Note 10**).
6. Calculate the proportions of different labeled components in the mixture using the Mascot server (Matrix Science) and Scaffold software (Proteome Software).
7. Based on the calculated proportions, equal amounts of peptides from different TMT-labeled samples were mixed and dried by vacuum centrifugation at 45 °C.

3.8 Gradient Elution Program

The gradient elution program was employed through a C18 stationary phase (2.1 mm, 100 mm, 2.6 m, Kinetex, Phenomenex, Cat. No. 00D-4462-AN). Fractionation was conducted using a High-Performance Liquid Chromatography (HPLC) system.

1. Prepare A-phase (HPLC-grade water) and B-phase (pure ACN).
2. Prepare C-phase (20 mM aqueous solution of ammonium acetate, pH 10.0) and D-phase (20 mM ACN solution of ammonium acetate, pH 10.0) (*see* **Note 11**).
3. Program the column rinse to infuse for 3 min at a 0.2 mL/min flow rate (*see* **Note 12**).
4. Adjust the phase proportions and rinsing times as specified in the following gradient to prepare for sample injection. Specifically, 30% A-phase, 70% B-phase, 10 min; 50% A-phase, 50% B-phase, 10 min; 70% A-phase, 30% B-phase, 10 min; 95% A-phase, 5% B-phase, 10 min.
5. The wash rate was reduced to zero, and the C-phase and D-phase were shifted with ammonium acetate, with an infusion time of 3 min.
6. Use the C-phase impact column with a flow rate of 0.2 mL/min for 95 min, combined with a Photo Diode Array (PDA).
7. The sample was dissolved in phase C (45 μL). A collecting plate is placed inside the instrument.
8. After gradient separation, each sample was pooled into 10 combined fractions (*see* **Note 13**).

3.9 LC-MS/MS Analysis

Analyze peptides by LC-MS/MS using a Q-Exactive high-resolution mass spectrometer (Thermo Fisher Scientific, Waltham, MA, USA) coupled with a nano-Acquity HPLC system (Waters, Milford, MA, USA). And the peptides were loaded onto a trap column (Acclaim PepMap, 75 μm × 2 cm, 3 μm, C18, 100 Å; Thermo Fisher Scientific, Waltham, MA, USA) and separated on an analytical column (Aqua, 100 μm × 15 cm, 3 μm, C18, 125 Å; Phenomenex, Los Angeles, CA, USA). The relevant Settings are described below.

1. Set the flow rate to 400 nL/min.
2. Elute peptides using a 125-min gradient with mobile A-phase (0.1% FA in water) and mobile B-phase (0.1% FA in ACN).
3. MS scans were acquired over an m/z range of 300–1800 with a resolution of 70,000.
4. Set the dynamic exclusion time to 20 s and the scan resolution for the top 10 most intense ions in the linear ion trap to 17,500.
5. Data analysis for identification and quantification was performed using the Mascot server and Scaffold software.
6. Remove common contaminants, decoys, and a Mann–Whitney U-test with Benjamini-Hochberg (BH) test $P < 0.05$, then remaining proteins were classified as "value detected."

3.10 Data Analysis

We will generally filter out candidate proteins with a normalized fold change (SGT1-HSP90/Citrine) of ≥1.5 from "value detected," representing high confidence in interaction with the bait proteins, SGT1-HSP90.

Interactions between candidate proteins of *N. benthamiana* were predicted using a corresponding subset of *A. thaliana* orthologs through the STRING online platform (https://cn.string-db.org/). The resulting data were visualized with Cytoscape software to construct the PPI network of SGT1-HSP90 (Fig. 2b).

4 Notes

1. Avoid long-term storage of the buffers and prepare them freshly each time.
2. The infiltration time and the final OD_{600} of *Agrobacterium* mixtures used to collect the leaf samples are according to the specific experimental conditions.
3. The leaf powder turns dark green when dissolved in RIPA extraction buffer. The samples were inverted multiple times to ensure protein extraction.
4. Use the pipette to diagonally insert the RIPA extraction buffer into the desalting columns to prevent them from skewing.
5. Mark the desalting columns to prevent confusion between samples.
6. Relevant protein extracts should be kept on ice or stored at 4 °C until subsequent experiments. Long-term storage "Input" needs to be frozen at −80 °C.
7. The bead volume should be controlled to avoid clogging the LC-MS/MS column. A gradient experiment to determine the optimal bead volume for each sample is advisable.
8. To resuspend the beads, invert the tubes or gently tap them with your fingers to mix evenly, instead of using a pipette, which minimizes beads loss. This protocol should be followed in all steps for resuspending the beads.
9. The TMTsixplex powder was first brought to room temperature (RT) and then dissolved in a constant-temperature oscillating metal bath by adding 120 μL of ACN, rather than using a pipette.
10. Based on quality control results, TMT labeling efficiency can be measured, and the subsequent blending ratio of samples can be calculated.
11. These solutions should be placed in the ultrasonic machine for over 20 min to eliminate air bubbles.

12. The C18 stationary phase is saved in 70% ACN.
13. Using gradient separation, one sample can be divided into 44 fractions, each containing varying concentrations of peptides. These 44 fractions should be combined into 10 fractions with equal peptide content for LC-MS/MS identification. The final result for the sample's LC-MS/MS identification will be derived from the combined data of the 10 fractions.

Acknowledgments

We thank Drs. Dawei Li and Meng Yang at China Agricultural University for their suggestions on this method. This work was supported by the National Natural Science Foundation of China (32320103003 and 32122070).

References

1. Yang X, Wen Z, Zhang D, Li Z, Li D, Nagalakshmi U et al (2021) Proximity labeling: an emerging tool for probing in planta molecular interactions. Plant Commun 2:100137
2. Branon TC, Bosch JA, Sanchez AD, Udeshi ND, Svinkina T, Carr SA et al (2018) Efficient proximity labeling in living cells and organisms with TurboID. Nat Biotechnol 36:880–887
3. Xu S-L, Shrestha R, Karunadasa SS, Xie P-Q (2023) Proximity labeling in plants. Annu Rev Plant Biol 74:285–312
4. Zhang Y, Song G, Lal NK, Nagalakshmi U, Li Y, Zheng W et al (2019) TurboID-based proximity labeling reveals that UBR7 is a regulator of N NLR immune receptor-mediated immunity. Nat Commun 10:3252
5. Mair A, Xu S-L, Branon TC, Ting AY, Bergmann DC (2019) Proximity labeling of protein complexes and cell-type-specific organellar proteomes in *Arabidopsis* enabled by TurboID. elife 8:e47864
6. Larochelle M, Bergeron D, Arcand B, Bachand F (2019) Proximity-dependent biotinylation mediated by TurboID to identify protein–protein interaction networks in yeast. J Cell Sci 132:jcs232249
7. Cho KF, Branon TC, Rajeev S, Svinkina T, Udeshi ND, Thoudam T et al (2020) Split-TurboID enables contact-dependent proximity labeling in cells. Proc Natl Acad Sci USA 117:12143–12154
8. Cho KF, Branon TC, Udeshi ND, Myers SA, Carr SA, Ting AY (2020) Proximity labeling in mammalian cells with TurboID and split-TurboID. Nat Protoc 15:3971–3999
9. Zhang D, Yang X, Wen Z, Li Z, Zhang X, Zhong C et al (2024) Proxitome profiling reveals a conserved SGT1-NSL1 signaling module that activates NLR-mediated immunity. Mol Plant 17:1369–1391
10. Tan H, Shi C, Macho AP, Lozano-Durán R (2024) The cartography of plant immunity: proximity labeling puts a novel SGT1–NSL1 regulatory module on the map. Mol Plant 17:1645–1647
11. Kruger NJ (2009) The Bradford method for protein quantitation. In: Walker JM (ed) The protein protocols handbook. Humana Press, Totowa, pp 17–24

Chapter 12

Virus-Induced Gene Silencing for Identifying Autoimmune Regulators in Arabidopsis

Suji Ye, Sung-Il Kim, and Ping He

Abstract

Virus-induced gene silencing (VIGS) has been applied as a functional genomics tool across diverse plant species. Integrated with the *Arabidopsis* sequence-tagged T-DNA homozygous mutant library, VIGS enables an efficient screening approach that combines features of both forward and reverse genetics, facilitating the identification of novel regulators in plant immunity. Plant defense against pathogens relies on a two-layered immune system, classified as pattern-triggered immunity (PTI) and effector-triggered immunity (ETI). Dysregulation of key PTI or ETI components can lead to excessive or uncontrolled cell death. The cell death phenotype offers a unique avenue for genetic screens aimed at identifying suppressors of immune-related cell death. However, conventional genetic approaches face limitations due to seedling lethality and the consequent lack of viable seeds, restricting their efficiency. Here, we describe an *Agrobacterium*-mediated transient VIGS assay optimized for systematic gene silencing at seedling stages, leading to cell death phenotypes. This method enables high-throughput screening for cell death suppressors using T-DNA homozygous mutant collections. The platform provides a rapid, cost-efficient strategy for uncovering key regulators of plant immune signaling, offering new insights into mechanisms governing immune homeostasis and cell death suppression.

Key words Virus-induced gene silencing (VIGS), Functional genomics, Plant immunity, Pattern-triggered immunity (PTI), Effector-triggered immunity (ETI), Cell death

1 Introduction

Virus-induced gene silencing (VIGS) is an RNA-mediated reverse genetic technique used for functional analysis of genes in plants. It is widely utilized to transiently downregulate endogenous gene expression through post-transcriptional gene silencing (PTGS) across various plant species, including tobacco, tomato, cotton, and *Arabidopsis* [1–4]. VIGS functions by delivering small RNA molecules into plant cells, thereby triggering the PTGS machinery and effectively suppressing target gene expression [5].

Among various VIGS systems, tobacco rattle virus (TRV)-based VIGS is extensively applied as a reverse genetic tool in dicot

Tsutomu Kawasaki and Hirofumi Yoshioka (eds.), *Plant Immunity: Methods and Protocols*, Methods in Molecular Biology, vol. 3012,
https://doi.org/10.1007/978-1-0716-5138-4_12,

model plants such as *Arabidopsis* and *Nicotiana* [6, 7]. TRV, a bipartite, positive-sense RNA virus, has a broad host range and infects over 400 plant species across multiple families, while typically causing only mild symptoms. It carries two genomic RNA segments, RNA1 and RNA2 [8]. TRV-RNA1 encodes an RNA-dependent RNA polymerase (RdRp), a movement protein (MP), and a cysteine-rich protein, all essential for replication and movement within the host. Meanwhile, TRV-RNA2 contains genes encoding for the coat protein (CP) and non-structural proteins that facilitate virion formation and transmission [1]. These two RNA segments, with modifications to reduce virulence symptoms, were cloned into broad-host-range binary vectors and can be delivered to plant cells via *Agrobacterium*-mediated transformation for TRV-mediated VIGS [9].

As a widely adopted functional genomics approach, VIGS has been prominently applied to investigate plant immune responses by transiently silencing genes involved in defense signaling pathways. Plants defend against pathogen attacks through a two-layered immune system [10, 11]. The first layer of immune responses is activated when plasma membrane-resident pattern-recognition receptors (PRRs) recognize pathogen/microbe-associated molecular patterns (PAMPs/MAMPs) and plant-derived damage-associated molecular patterns (DAMPs) or phytocytokines, mounting PTI [12–14]. Typical PTI-mediated signaling responses include rapid mitogen-activated protein kinase (MAPK) activation, reactive oxygen species (ROS) burst, an increase in cytosolic calcium concentration, and gene transcriptional reprogramming [15, 16]. PTI provides basal defense for plants to fend off a broad range of non-adapted pathogens. However, some adapted pathogens deploy virulence strategies, including delivering effector proteins into plant cells, to interfere with the host physiology or inhibit PTI, leading to effector-triggered susceptibility (ETS) [10, 11]. To counteract this, plants have evolved a second layer of the immune system, termed effector-triggered immunity (ETI), which is triggered by recognition of effector by nucleotide-binding domain leucine-rich repeat-containing receptors (NLRs) to combat pathogen virulence [11]. ETI is often associated with hypersensitive response (HR) that manifests localized cell death to limit pathogen spread [10, 11].

Although classified as a two-layered immune system, PTI and ETI intertwine at multiple levels with mutual potentiation for robust immunity [17]. PTI also constrains ETI, likely preventing strong immune activation-associated cell death and autoimmunity, thereby attributing balanced immune homeostasis [18, 19]. A long-standing observation is that alteration of the PRR-activated MAPK cascade, consisting of sequentially phosphorylated MEKK1-MKK1/2-MPK4, leads to strong immunity and cell death due to activation of NLR SUMM2 (SUPPRESSOR OF *mkk1mkk2* 2), supporting that the PRR-mediated MAPK cascade can suppress NLR-mediated ETI cell death activation [20–22]. Another well-established example of PTI constraining ETI is the disruption of BRASSINOSTEROID INSENSITIVE 1-ASSOCIATED RECEPTOR KINASE 1/SOMATIC EMBRYOGENESIS RECEPTOR KINASE 4 (BAK1/SERK4), shared co-receptors for multiple PRRs crucial for PTI signaling initiation, *which* leads to strong autoimmunity and cell death [23, 24]. This cell death has recently been indicated to be associated with NLR activation, further elaborating on PTI suppressing ETI-associated strong immunity and cell death at the receptor level [25–27]. Although various components related to cell death and immune homeostasis regulation are emerging, many unknown aspects and signaling mechanisms remain.

VIGS offers several advantages for genetic screens and functional analysis of autoimmune and cell death regulators. First, VIGS is applied post-germination at seedling stages and induces systemic and potent PTGS-mediated gene silencing, thereby bypassing autoimmune-triggered embryonic and seedling lethality [6, 28]. This approach is particularly valuable when studying immune regulators involved in autoimmunity, as knocking out these genes often leads to embryonic or seedling lethality due to constitutive activation of defense signaling, as observed in the MEKK1-, MKK1/MKK2-, and MPK4-mediated MAPK cascade pathway [28–31]. Second, VIGS vectors can be designed to target multiple genes of interest simultaneously, thereby efficiently silencing functionally redundant genes at once. For instance, BAK1 and SERK4 function redundantly in negatively regulating autoimmunity as the only strong allele of double mutants *bak1-4/serk4-1*, but not the individual cognate single mutants, showed seedling lethality [23, 24]. The silencing of BAK1 and SERK4 in one VIGS vector leads to strong cell death [6]. Third, VIGS can be readily applied to a collection of mutants to evaluate the specific involvement of gene family members. This is exemplified by the specific involvement of endoplasmic reticulum quality control (ERQC) components, cyclic nucleotide-gated ion channel (CNGC) family members, and receptor kinases in BAK1/SERK4-mediated autoimmune regulation [6, 27, 32]. Finally, when integrated with the *Arabidopsis* sequence-tagged T-DNA homozygous mutant library, VIGS is a powerful high-throughput screen approach, combining the key elements of both forward and reverse genetics approaches [6, 28]. Initially, this method allows an unbiased forward genetic screen from phenotypes to mutants, thus leading to the uncovering of previously uncharacterized mutants. Once the mutant is identified, the putative causal mutation can be revealed instantaneously since the mutant collection is a sequence-tagged T-DNA homozygous mutant library. This also allows rapid validation of causal mutations using additional alleles, a hallmark of reverse genetics.

Thus, VIGS empowers fast and high-throughput genetic screening. Here, we describe the detailed procedure for establishing an efficient VIGS assay and carrying out genetic screening for cell death suppressors.

2 Materials

2.1 Plants and Growth Conditions

1. Plant materials: Two-week-old soil-grown wild-type (WT) *Arabidopsis thaliana* ecotype Col-0, T-DNA insertion mutant library obtained from Arabidopsis Biological Resource Center (https://abrc.osu.edu/), such as stock number CS27941: One-allele set #1 of 6866 confirmed T-DNA insertion lines, each line representing an insertion in a different locus, or a customized list of T-DNA insertion lines (*see* **Note 1**).
2. Plant growth condition: 23 °C, 50% relative humidity, 120 $\mu Em^{-2}\ s^{-1}$ light, 12-h light/12-h dark photoperiod (*see* **Note 2**).
3. Soil: Propagation mix (Sungro horticulture sunshine® mix #5, USA).

2.2 Cloning VIGS Construction

1. *Escherichia coli* MC1061 competent cells stored in 10% glycerol at −80 °C.
2. *Agrobacterium tumefaciens* GV3101 competent cells stored in 10% glycerol at −80 °C.
3. TRIzol™ Reagent (Invitrogen, USA).
4. Chloroform.
5. 10× DNase I reaction buffer.
6. 2 U/μL DNase I.
7. 40 mM oligo dT.
8. 2.5 mM dNTP.
9. 10× M-MuLV Reverse Transcriptase Reaction Buffer (New England BioLabs, USA).
10. M-MulV Reverse Transcriptase (200 U/μL, New England BioLabs, USA).
11. RNase inhibitor, Murine (40 U/μL, New England BioLabs, USA).
12. PCR amplification regents: 10× reaction buffer, 10 mM dNTP, and *Taq* DNA Polymerase (New England BioLabs, USA).
13. Restriction enzymes: EcoRI and KpnI, rCutSmart™ Buffer (New England BioLabs, USA).

14. DNA ligation kit: 10× T4 DNA ligase buffer and T4 DNA ligase (4 U/μL) (New England BioLabs, USA).
15. pYL156 (pTRV2-RNA2) vector.
16. Luria-Bertani (LB) plates containing 50 μg/mL Kanamycin.
17. LB liquid medium containing 50 μg/mL Kanamycin and 50 μg/mL Gentamycin.

2.3 Agrobacterium Infiltration

1. Agroinfiltration solution: 10 mM $MgCl_2$, 10 mM MES [2-(N-morpholino) ethanesulfonic acid, pH 5.6], and 200 mM acetosyringone (3′,5′-Dimethoxy-4′--hydroxyacetophenone) (*see* **Note 3**).
2. RT-qPCR reagent: 2× iQ™ SYBR Green Supermix (Bio-Rad Laboratories, Inc., USA).
3. 4× SDS loading buffer: 0.25 M Tris-Cl, pH 6.8, 4% SDS, 0.1% bromophenol blue, 40% glycerol, 4% β-mercaptoethanol.
4. 30% Acrylamide-Bis (Bio-Rad Laboratories, Inc., USA).
5. 1 M Tris–HCl, pH 8.8.
6. 1 M Tris–HCl, pH 6.8.
7. 20% SDS.
8. 10% APS (ammonium persulfate).
9. TEMED (Bio-Rad Laboratories, Inc., USA).
10. 1× Tris-glycine-SDS running buffer (dilute 10× solution: 121 g Tris, 567 g glycine, and 40 g SDS in 4 L ddH_2O).
11. 1× Transfer buffer (10% of 10× Transfer buffer, 5% methanol, 1% SDS).
12. 0.45 μm PVDF membrane (Bio-Rad Laboratories, Inc., USA).
13. Filter paper.
14. 5% skim milk in PBST (PBS with 0.1% Tween 20).
15. Polyclonal anti-pathogenesis-related protein 1 (PR1) antibody (Agrisera Antibodies, SWEDEN).
16. Anti-rabbit IgG HRP-conjugated secondary antibody.
17. Clarity Western ECL Substrate (Bio-Rad Laboratories, Inc., USA).

2.4 Equipment

1. 37 °C and 28 °C incubators (VWR Gravity Convection Incubator, USA).
2. Incubating shaker (INFORS HT Multitron Standard 2.0 Incubator, USA).
3. 1-mL disposable syringes without needles (BD 1 mL tuberculin syringe without needle, USA).
4. PCR machine (Eppendorf Mastercycler X50S, USA).

5. Hole puncher (0.5 mm diameter).
6. TissueLyser (QIAGEN TissueLyser II sample disruption with 24-well adapters, USA).

3 Methods

3.1 Grow *Arabidopsis* Plants

1. Soak seeds (WT and T-DNA mutant seeds from ABRC) in water for stratification to synchronize germinations at 4 °C for 2–4 days (*see* **Note 4**).
2. Fill the soil in square pots (12.5 cm × 12.5 cm) in a tray and soak the potting soil by watering in the tray to ensure thorough saturation; sow seeds in the soil at a density of 16 seeds per pot in a 4 × 4 arrangement (*see* **Note 5**).
3. Cover the trays with the transparent dome and place plants in the controlled growth chamber or rooms (23 °C, 50% relative humidity, 120 μEm^{-2} s^{-1} light, 12-h light/12-h dark photoperiod) until two cotyledons start to emerge (about 1 week) and remove the dome until the two cotyledons are fully expanded (*see* **Note 6**).

3.2 Cloning the Gene-of-Interest (GOI) into the pTRV-RNA2 Vector

1. Homogenize 50–100 mg of plant tissues of WT in 1 mL of TRIzol™ Reagent (Invitrogen, USA) to isolate total RNA (*see* **Note** 7).
2. Add 200 μL of chloroform and vortex thoroughly to mix.
3. Centrifuge at 14,462 g for 10 min at 4 °C.
4. Transfer the supernatant (aqueous phase) to a new 1.5 mL microtube.
5. Add 0.5 mL of isopropanol and then centrifuge at 14,462 g for 10 mins at 4 °C.
6. Gently invert the pellet in 75% ethanol, then discard the supernatant.
7. Air-dry the pellet and dissolve the total RNA in 50 μL of DEPC-dH_2O.
8. Mix 1 μg of total RNA (500–1000 ng/μL) with 0.5 μL 10× DNase buffer and 1 μL DNase in a total volume of 5 μL.
9. Incubate at 37 °C for 15 min and then heat at 75 °C for 10 min and chill at 4 °C.
10. Add 1 μL of 40 μM oligo (dT) and 2 μL of 2.5 mM dNTP to the DNase-treated RNA (total volume 8 μL).
11. Heat 75 °C for 5 min and chill at 4 °C.
12. Synthesize cDNA with a reverse transcription cocktail (2 μL of 10× M-MuLV reverse transcriptase reaction buffer, 0.5 μL of

RNase inhibitor, 1 μL M-MuLV reverse transcriptase, and 8.5 μL of dH_2O in a total volume of 20 μL).

13. Incubate at 42 °C for 1 h and then heat at 75 °C for 15 min.
14. Chill at 4 °C and dilute 20-fold of volume with dH_2O for PCR.
15. Select a 300–500 bp target region of the GOI and design a pair of primers incorporating EcoRI at the 5′ end and KpnI at the 3′ end. For example, the forward primer should begin with the EcoRI recognition sequence (5′ GAATTC) followed by 18–25 bp of your target sequence. Similarly, the reverse primer should start with the KpnI recognition sequence (5′ GGTAC) followed by 18–25 bp corresponding to the reverse complementary sequence of your target (*see* **Notes 8** and **9**).
16. Amplify GOI by PCR with *Taq* DNA polymerase from the above cDNA library (*see* **Note 10**).
17. Precipitate PCR products with ethanol and wash with 75% ethanol twice to remove salts followed by digestion with EcoRI and KpnI. Simultaneously, digest the pTRV-RNA2 vector (pYL156) with the same restriction enzymes EcoRI and KpnI.
18. Ligate the digested PCR products and pYL156 vector in a 20 μL total volume containing 1 μL of vector DNA (5–10 ng), 5 μL of insert DNA (20–50 ng), 2 μL of 10× T4 DNA ligase buffer, and 0.2 μL of T4 DNA ligase (4 U/μL) at room temperature for approximately 2 h.
19. Add 10 μL of ligation reaction sample to 50 μL *E. coli* competent cells and heat shock at 42 °C for 1 min. Then chill on ice and recover with 1 mL LB medium at 37 °C using a shaking incubator or roller drum for 40 min.
20. Spread the transformed *E. coli* on an LB plate supplemented with Kanamycin (50 μg/mL) and incubate at 37 °C overnight.
21. Pick about ten single colonies for miniprep plasmid DNA isolation and digest plasmid with EcoRI and KpnI, and then analyze the digested products by agarose gel electrophoresis to confirm the presence of the insert at the expected size.
22. The colonies will be further confirmed by Sanger Sequencing and further transformed into *Agrobacterium tumefaciens* GV3101 using the heat shock method. Mix 4 μL of plasmid DNA with 100 μL of competent *Agrobacterium* cells, incubate on ice for 5 min, subject to liquid nitrogen freezing for 5 min, and then incubate in a 37 °C water bath for 5 min. The transformed competent cells are recovered with 1 mL LB medium at 28 °C incubator with shaking for 2 h.

23. Spread the culture on an LB plate containing Kanamycin (50 μg/mL) and Gentamycin (50 μg/mL) and incubate the plate at 28 °C for 2–3 days.

3.3 Agrobacterium Infiltration and VIGS Assays in WT Plants

1. Select a single colony of *Agrobacterium tumefaciens* carrying pYL156-TRV1, pYL156-TRV2-*GFP*, pYL156-TRV2-*CLA1*, and pYL156-TRV2-*GOI* from an LB agar plate containing Kanamycin and Gentamycin (*see* **Note 11**).
2. Culture the selected colonies in 2 mL of LB liquid medium containing Kanamycin (50 mg/mL) and Gentamycin (50 mg/mL) in a roller drum at 28 °C overnight (*see* **Notes 12** and **13**).
3. Transfer 1 mL of *Agrobacterium* culture into 10 mL of LB liquid containing Kanamycin (25 mg/mL) and Gentamycin (20 mg/mL) in a 50 mL-Erlenmeyer flask and incubate the culture overnight at 28 °C in a shaker at 200 rpm.
4. Centrifuge the bacterial culture at 3220 g for 10 min to harvest cells, then resuspend the pellet in *Agrobacterium* infiltration solution after adjusting the $OD_{600} = 1.5$ each (*see* **Notes 14** and **15**).
5. Leave the bacterial culture at room temperature for 3 h (*see* **Note 16**).
6. Mix the Agrobacterial culture suspension of pYL156-TRV1 with pYL156-TRV2-*GFP*, pYL156-TRV2-*CLA1*, or pYL156-TRV2-*GOI* at a 1:1 ratio, respectively (Fig. 1a).

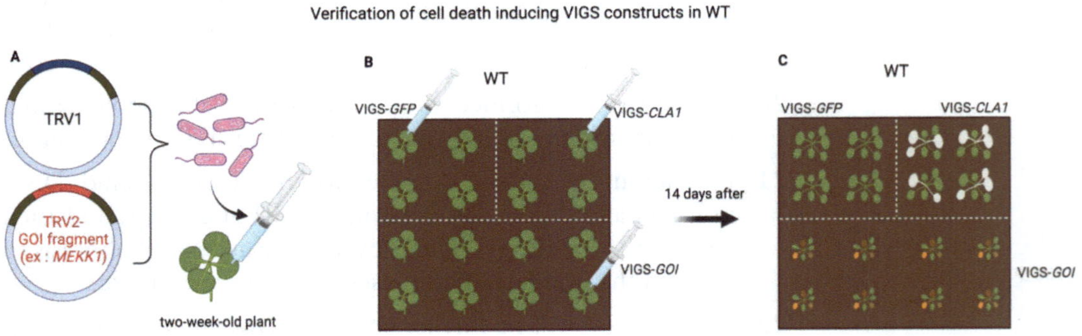

Fig. 1 Workflow of VIGS-mediated gene silencing and phenotype assessment. (**a**) Two VIGS plasmids: pTRV1, containing the *RNA1* operon (encoding RNA-dependent RNA polymerase, movement protein, and cysteine-rich protein), and pTRV2, containing the *RNA2* operon (encoding two coat proteins as sub-genomic RNAs and the gene of interest [GOI]), were separately transformed into *Agrobacterium*. The first pair of true leaves of wild-type (WT) were co-infiltrated with a mixture of *Agrobacterium* strains harboring pTRV1 and pTRV2-*GOI*. (**b**) Three VIGS constructs—technical control (VIGS-*CLA1*), experimental control (VIGS-*GFP*), and knockdown target (VIGS-*GOI*)—were infiltrated into WT plants grown in the same pot under identical conditions. (**c**) VIGS-induced phenotypes were evaluated 14 days post-infiltration (dpi). VIGS-*CLA1* (positive control) induces an albino phenotype, while VIGS-*GFP* (negative control) shows normal growth. The VIGS-*GOI* construct was assessed for target-specific phenotypes

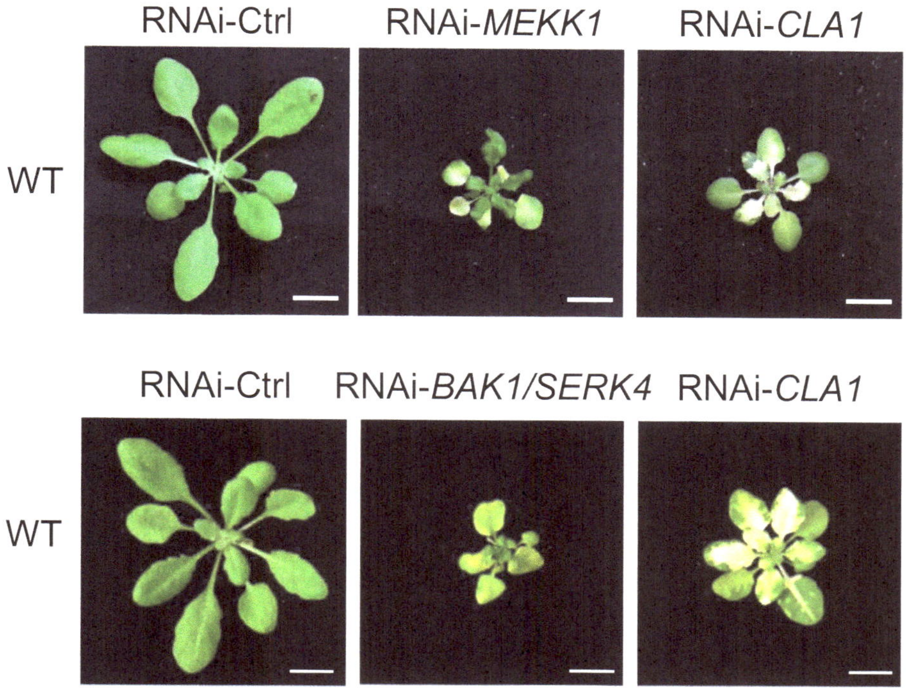

Fig. 2 VIGS-*MEKK1* and VIGS-*BAK1/SERK4* cause cell death in WT. A practical application of the schematic presented in Fig. 1 with constructs targeting *MEKK1* and *BAK1/SERK4* exhibiting strong dwarfism and growth defects. The *mekk1* single mutant and *bak1-4/serk4-1* double null mutants are seedling lethal. To silence *MEKK1* and *BAK1/SERK4*, pTRV1 along with either pTRV2-*MEKK1* or pTRV2-*BAK1/SERK4* is infiltrated into 2-week-old WT plants. Phenotype is observed at 14 days post-infiltration. The plant infiltrated with VIGS-*GFP* displays normal growth, and those with VIGS-*CLA1* exhibit an albino phenotype. In contrast, VIGS-*MEKK1* (top) and VIGS-*BAK1/SERK4* (bottom) result in dwarfism and growth defects with chlorosis leaves. The white scale bar indicates 1 cm

7. Manually infiltrate the Agrobacterial mixture carrying pYL156-TRV1 paired with either pYL156-TRV2-*GFP*, pYL156-TRV2-*CLA1*, or pYL156-TRV2-*GOI* into the abaxial of cotyledons of *Arabidopsis* WT using a 1-mL needleless syringe (Fig. 1b) (*see* **Notes 17** and **18**).
8. Cover the infiltrated plants with a dome overnight in a growth chamber (*see* **Note 19**).
9. Remove the dome and continue growing the plants under the same conditions: 23 °C, 50% relative humidity, 120 μEm^{-2} s^{-1} light, 12-h light/12-h dark.
10. Approximately 10 days after infiltration, the albino phenotype in VIGS-*CLA1* silencing plants begins to appear, which serves as a visual marker for VIGS efficiency. By around 14 days, VIGS-*GOI* could start to show a phenotype. For example, compared to the VIGS-*GFP* plants, VIGS-*MEKK1* plants become dwarf with older leaves showing chlorotic and younger true leaves exhibiting dark green and twisted, and VIGS-

BAK1/SERK4 plants also become severely dwarf with older leaves showing chlorotic and younger true leaves exhibiting rounder in shape and dark green, with difficulties to emerge (Figs. 1c and 2) (*see* **Note 20**).

3.4 Verification of Gene Silencing Efficiency

1. Homogenize 50–100 mg of leaf tissue of VIGS-*GFP* and VIGS-*GOI* in 1 mL of TRIzol™ Reagent to isolate total RNA.
2. Extract RNA for cDNA synthesis as described in Subheading 3.2; dilute 20 to 50-fold of volume with dH_2O.
3. Take 4.5 μL of cDNA and mix with 5 μL of 2× SYBR Green Master Mix and 0.5 μL of mixed forward and reverse primers (10 μM each) (*see* **Note 21**).
4. Perform RT-PCR or qPCR using target and housekeeping gene-specific primers. Then, carry out quantitative analysis between the control and target genes, or calculate the relative transcript level from the difference between the threshold cycle (Ct) values of reference and target genes, respectively.

3.5 Screening for Cell Death Suppressors Triggered by VIGS-GOI in Arabidopsis T-DNA Mutant Library

1. If VIGS-*GOI* triggers cell death in WT plants, obtain the T-DNA insertion mutant library from ABRC or compile a list of T-DNA mutant lines to screen potential cell death suppressors in response to VIGS-*GOI* (*see* **Note 22**).
2. Perform the infiltration process using the *Agrobacterium* culture and screen the selected T-DNA mutant lines (*see* **Note 23**).
3. Compare the results to VIGS-*GOI* in WT to identify any mutant lines that either suppress or enhance cell death triggered by VIGS-*GOI* (Fig. 3) (*see* **Note 24**).
4. Repeat the screening of promising candidate mutant lines, including VIGS-*GFP* as a vector control and VIGS-*CLA1* as a visual marker for VIGS efficiency. Different mutants show various degrees of cell death suppression activities. For example, *summ2-8* and *mekk2* mutants suppress VIGS-*MEKK1* cell death, whereas *btl2-1* and *cngc19/20* mutants suppress VIGS-*BAK1/SERK4* cell death (Fig. 4). The VIGS-*GFP* plants could indicate whether the mutant line has any growth defects, and VIGS-*CLA1* plants could suggest whether the mutant line interferes with the gene silencing machinery.
5. Identify the gene linked with the T-DNA insertion from the mutant line based on SALK annotation (https://www.arabidopsis.org/). Order additional mutant alleles for the same gene to repeat the VIGS-*GFP*, VIGS-*GOI*, and VIGS-*CLA1* inoculation to confirm that cell death suppression is linked to the gene (*see* **Note 21**).

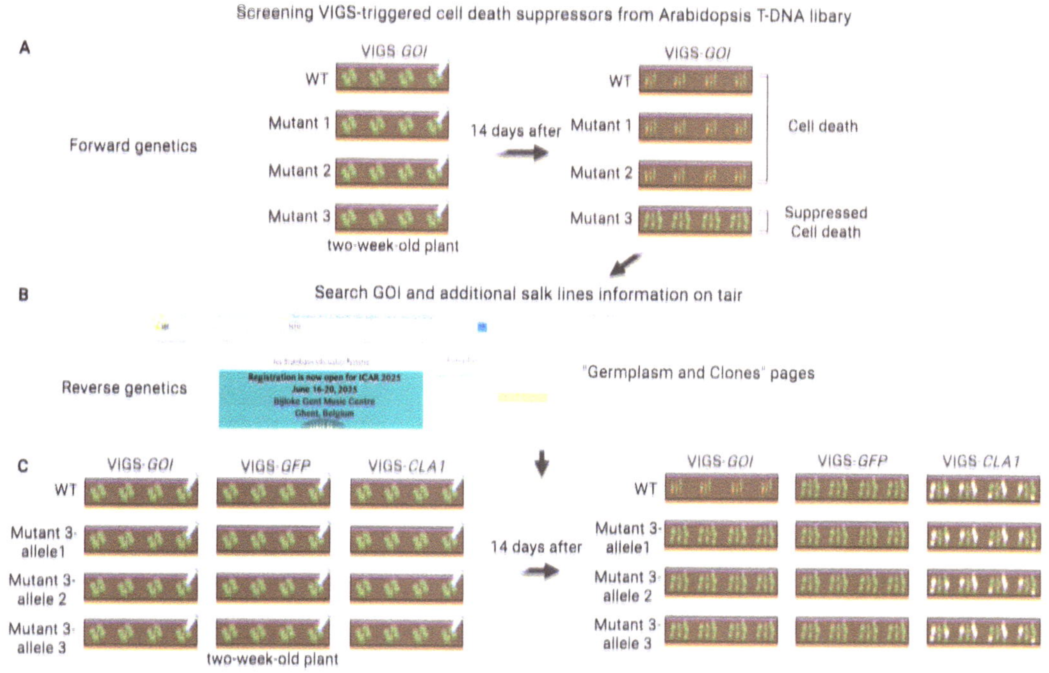

Fig. 3 Workflow of screen for cell death suppressors with the *Arabidopsis* T-DNA library. Suppressors of VIGS-triggered cell death can be identified using the *Arabidopsis* T-DNA library. To achieve this, pTRV1 and pTRV2-*GOI*, which have been verified to induce cell death phenotype in WT, are infiltrated into individual mutant lines of the T-DNA library. Among the candidates, if a T-DNA line infiltrated with VIGS-*GOI* displays a normal phenotype like mutant 3, in contrast to other lines exhibiting a cell death phenotype like mutant 1 and mutant 2, it can be considered a potential suppressor of VIGS-*GOI*-mediated cell death (**a**) Next, additional genetic alleles of the candidate mutant should be selected to confirm whether the causal mutation is linked to the suppressor function of GOI (**b**, **c**) VIGS-*GFP* and VIGS-*CLA1* are essential to rule out the possibility that the suppressed phenotype is caused by the infiltration process itself or by a generic effect from loss-of-function in PTGS machinery, respectively. This suggests that the gene disrupted in the T-DNA mutant line is likely involved in the *GOI*-mediated cell death pathway. *CLA1* Cloroplastos alterados 1

3.6 Western Blot to Test the PR1 Protein Level

1. Harvest two leaf discs (size: 0.5 cm diameter) from each plant in a 1.5-mL microtube and store them in liquid nitrogen.
2. Grind leaves 20 s (30 Hz) with TissueLyser by adding 2× SDS loading buffer and boil at 95 °C for 5 min.
3. Load 8–10 μL of each sample into the wells of a 12.5% SDS-PAGE.
4. Run the gel at 50–90 V through the stacking layer for 30–40 min, then increase to 90–130 V for the separating gel until clear separation of the protein marker is observed (*see* **Note 25**).

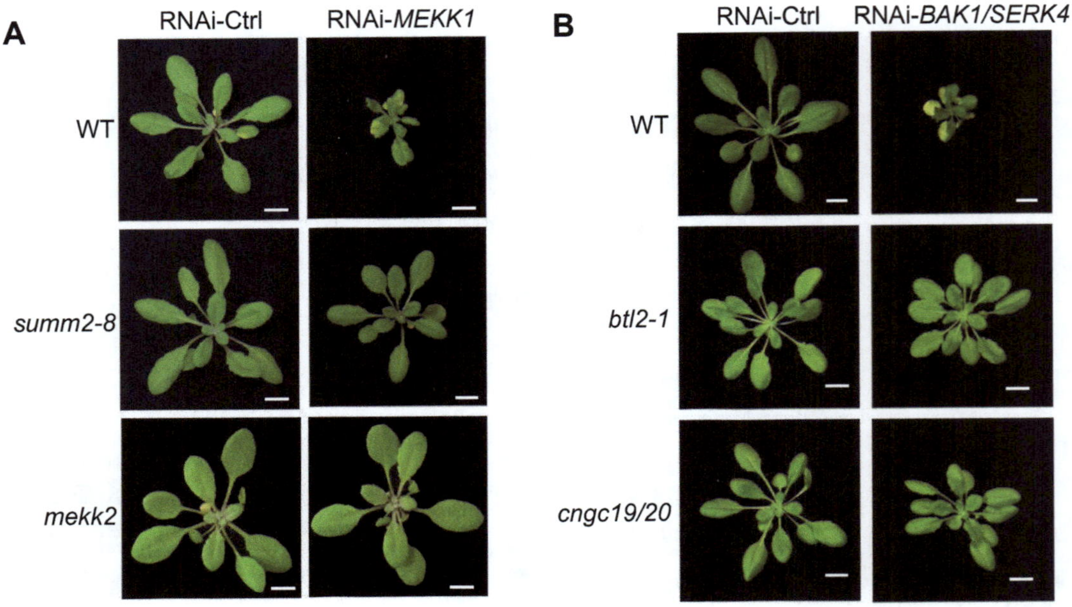

Fig. 4 Suppressors of VIGS-*MEKK1*- or VIGS-*BAK1/SERK4*-induced cell death. A practical application of the schematic in Fig. 3 using VIGS-*MEKK1* and VIGS-*BAK1/SERK4* to identify suppressors of cell death. pTRV1 and pTRV2-*MEKK1* are infiltrated into two-week-old *Arabidopsis* WT, *summ2-8* and *mekk2* mutants, while pTRV1 and pTRV2-*BAK1/SERK4* are infiltrated into WT, *btl2-1,* and *cngc19/20* mutants. *SUMM2* and *MEKK2* are key components of the *mekk1*-mediated cell death pathway, while *BTL2* and *CNGC19/20* are involved in the *BAK1/SERK4*-mediated cell death pathway. Phenotypic evaluation is conducted at 14 days post-infiltration. VIGS-*MEKK1*-induced cell death was suppressed in *summ2-8* and *mekk2*, and VIGS-*BAK1/SERK4*-induced cell death was suppressed in the *btl2* and *cngc19/20* mutants, highlighting the involvement of these genes in their respective pathways. *BAK1/SERK4*: BRASSINOSTEROID INSENSITIVE 1-ASSOCIAATED RECEPTOR KINASE 1/SOMATIC EMBRYOGENESIS RECEPTOR KINASE 4; *SUMM2*: SUPPRESSOR OF *mkk1mkk2* 2; *MEKK2*: MITOGEN-ACTIVATED PROTEIN KINASE KINASE KINASE 2; *BTL2*: BAK-TO-LIFE 2; and *CNGC19/20*: CYCLIC NUCLEOTIDE-GATED ION CHANNEL 19/20. The white scale bar indicates 1 cm

5. Transfer proteins onto the PVDF membrane and block the membrane with 5% milk in PBST for 40–60 min at room temperature.
6. Incubate with anti-PR1 antibody (1:2000 dilution) in 5% skim milk in PBST overnight at 4 °C on a shaker.
7. Wash the membranes four times with 1× PBST for 10 min each.
8. Incubate the membrane with HRP-conjugated anti-rabbit IgG antibody (1:10,000 dilution) in 5% skim milk in PBST for 1 h at RT with shaking.
9. Wash the membrane three times with 1× PBST for 10 min each.
10. Develop with the ECL (Enhanced Chemiluminescence) method.

4 Notes

1. The optimal stage for performing VIGS is when the first pair of true leaves is similar in size to the cotyledons, typically around 14 days after germination. Silencing efficiency decreases once the second pair of true leaves emerges.
2. Maintaining stable growth conditions is crucial for achieving consistent and uniform silencing efficiency.
3. Although silwet, a nonionic organosilicon surfactant, is commonly used in *Agrobacterium*-mediated infiltration to enhance solution spreading and adhesion to hydrophobic plant surfaces, it may cause cell death in two-week-old *Arabidopsis* leaves. Thus, silwet is not recommended.
4. This could improve plant germination efficiency and uniformity, and eliminate thrips eggs.
5. After fully soaking the soil, remove excess water or replace it with a dry tray. Two to three seeds per spot are enough (depending on the germination rate), and then thin the seedlings after about 10 days, leaving one seedling per spot.
6. High humidity is crucial for germination; however, keeping the dome covered for over 10 days may hinder photosynthesis and cause hypocotyl elongation, resulting in unhealthy plants.
7. When starting with a large amount of tissues, the TRIzol method typically yields more RNA, but the overall quality is often compromised due to contaminants or incomplete phase separation.
8. For multiple targets, a fragment length of 200–300 bp is sufficient for each target. However, fragments shorter than 200 bp may result in reduced silencing efficiency.
9. Software like the VIGS Tool at Sol Genomics (https://vigs.solgenomics.net/) is helpful to select the best candidates based on off-target probability.
10. Compared to the vector containing a partial GFP sequence, the empty vector can induce plant growth defects resembling TRV infection phenotypes. Thus, it is recommended to use the vector carrying a partial *GFP* or *GUS* sequence as a control.
11. The albino phenotype of VIGS-*CLA1* confirms the correct procedure of VIGS and VIGS efficiency. More than 80% of VIGS-*CLA1* plants should show an albino phenotype.
12. It is highly recommended to select a single colony for experiments to secure clone identity.
13. Avoid using *Agrobacterium* from agar plates older than 1 month as viability may decrease leading to reduced functionality.

14. Keep in mind that the *Agrobacterial* culture carrying TRV1 should be prepared much more than the *Agrobacterial* culture carrying TRV2, as it needs to be mixed with each TRV2.
15. A 15 mL culture is sufficient for infiltrating 50–60 plants.
16. Leaving the bacteria suspended with infiltration solution at room temperature for 3 h can enhance silencing efficiency. It can be extended up to 24 h.
17. Avoid infiltrating the same leaf more than once to reduce physical damage.
18. Thoroughly remove excess bacteria from the leaf using Kimwipes after infiltration. Excessive bacteria on the leaf can obstruct respiration leading to impaired recovery post-infiltration, ultimately resulting in plant death.
19. The optimal temperature for *Agrobacterium* infection is around 22 °C, which could increase silencing efficiency.
20. If an obvious VIGS-triggered cell death phenotype is not observed, consider using VIGS constructs with different target fragments or multiple targets from the same *GOI*.
21. Try to design primers outside of the fragment for VIGS for RT-qPCR analysis.
22. Gene information can be found on the Arabidopsis Information Resource (TAIR, http://arabidopsis.org), where you can also select T-DNA library lines (TAIR → Search gene ID → Click 'Germplasm and Clones' tab). Information on T-DNA insertion locations and homozygous line numbers is available on the same platform. T-DNA insertion lines are available through ABRC (Arabidopsis Biological Resource Center, https://abrc.osu.edu/).
23. At this step, there is no need for VIGS-*GFP* and VIGS-*CLA1* controls; therefore, all plants can be used for VIGS-*GOI*, and eight plants per T-DNA mutant line are sufficient.
24. Promising T-DNA lines may show partial or full suppression of VIGS-triggered cell death when the *GOI* is silenced, whereas WT typically exhibits a cell death phenotype under VIGS-*GOI*.
25. Applying low voltage during the stacking phase helps ensure that all samples migrate uniformly and align at the same starting point.

Acknowledgments

We thank Drs. Libo Shan, Qiaochu Shen, and Lingxiao Ji for their helpful reading and feedback on the manuscript. Their comments and suggestions greatly improved the clarity of this work.

References

1. Liu Y, Schiff M, Dinesh-Kumar SP (2002) Virus-induced gene silencing in tomato. Plant J 31:777–786
2. Robertson D (2004) VIGS vectors for gene silencing: many targets, many tools. Annu Rev Plant Biol 55:495–519
3. Gao X, Britt JRC, Shan L et al (2011) Agrobacterium-mediated virus-induced gene silencing assay in cotton. J Vis Exp 54:2938
4. Li M, Li F, He P (2022) Plant gene silencing. Springer US, New York
5. Rössner C, Lotz D, Becker A (2022) VIGS goes viral: how VIGS transforms our understanding of plant science. Annu Rev Plant Biol 73:703–728
6. De Oliveira MVV, Xu G, Li B et al (2016) Specific control of Arabidopsis BAK1/SERK4-regulated cell death by protein glycosylation. Nat Plants 2:15218
7. Todd AT, Liu E, Polvi SL et al (2010) A functional genomics screen identifies diverse transcription factors that regulate alkaloid biosynthesis in Nicotiana benthamiana. Plant J 62:589–600
8. Macfarlane SA (2010) Tobraviruses - plant pathogens and tools for biotechnology. Mol Plant Pathol 11:577–583
9. Liu Y, Schiff M, Marathe R et al (2002) Tobacco Rar1, EDS1 and NPR1/NIM1 like genes are required for N-mediated resistance to tobacco mosaic virus. Plant J 30:415–429
10. Zhou J-M, Zhang Y (2020) Plant immunity: danger perception and signaling. Cell 181: 978–989
11. Jones JDG, Staskawicz BJ, Dangl JL (2024) The plant immune system: from discovery to deployment. Cell 187:2095–2116
12. Zhang C, Xie Y, He P et al (2024) Unlocking nature's defense: plant pattern recognition receptors as guardians against pathogenic threats. Mol Plant Microbe Interact 37:73–83
13. Escocard de Azevedo Manhães AM, Ortiz-Morea FA, He P et al (2021) Plant plasma membrane-resident receptors: surveillance for infections and coordination for growth and development. J Integr Plant Biol 63:79–101
14. Jian Y, Liu Z, He P et al (2024) An emerging connected view: phytocytokines in regulating stomatal, apoplastic, and vascular immunity. Curr Opin Plant Biol 82:102623
15. Snoeck S, Johanndrees O, Nürnberger T et al (2025) Plant pattern recognition receptors: from evolutionary insight to engineering. Nat Rev Genet 26:268–278
16. Yu X, Feng B, He P et al (2017) From chaos to harmony: responses and signaling upon microbial pattern recognition. Annu Rev Phytopathol 55:109–137
17. Yuan M, Ngou BPM, Ding P et al (2021) PTI-ETI crosstalk: an integrative view of plant immunity. Curr Opin Plant Biol 62:102030
18. Sun T, Zhang Y (2022) MAP kinase cascades in plant development and immune signaling. EMBO Rep 23:e53817
19. Ortiz-Morea FA, Liu J, Shan L et al (2021) Malectin-like receptor kinases as protector deities in plant immunity. Nat Plants 8:27–37
20. Zhang Z, Liu Y, Huang H et al (2017) The NLR protein SUMM2 senses the disruption of an immune signaling MAP kinase cascade via CRCK 3. EMBO Rep 18:292–302
21. Kong Q, Qu N, Gao M et al (2012) The MEKK1-MKK1/MKK2-MPK4 kinase cascade negatively regulates immunity mediated by a mitogen-activated protein kinase kinase kinase in *Arabidopsis*. Plant Cell 24:2225–2236
22. Zhang Z, Wu Y, Gao M et al (2012) Disruption of PAMP-induced MAP kinase cascade by a pseudomonas syring effector activates plant immunity mediated by the NB-LRR protein SUMM2. Cell Host Microbe 11:253–263
23. He K, Gou X, Yuan T et al (2007) BAK1 and BKK1 regulate brassinosteroid-dependent growth and brassinosteroid-independent cell-death pathways. Curr Biol 17:1109–1115
24. Kemmerling B, Schwedt A, Rodriguez P et al (2007) The BRI1-associated kinase 1, BAK1, has a brassinolide-independent role in plant cell-death control. Curr Biol 17:1116–1122
25. Schulze S, Yu L, Hua C et al (2022) The Arabidopsis TIR-NBS-LRR protein CSA1 guards BAK1-BIR3 homeostasis and mediates convergence of pattern- and effector-induced immune responses. Cell Host Microbe 30: 1717–1731.e6
26. Yang Y, Kim NH, Cevik V et al (2022) Allelic variation in the Arabidopsis TNL CHS3/CSA1 immune receptor pair reveals two functional cell-death regulatory modes. Cell Host Microbe 30:1701–1716.e5
27. Yu X, Xie Y, Luo D et al (2023) A phospho-switch constrains BTL2-mediated phytocytokine signaling in plant immunity. Cell 186: 2329–2344.e20
28. Yang Y, Liu J, Yin C et al (2020) RNA interference-based screen reveals concerted functions of MEKK2 and CRCK3 in plant cell death regulation. Plant Physiol 183:331–344

29. Liu J, Huang Y, Kong L et al (2020) The malectin-like receptor-like kinase LETUM1 modulates NLR protein SUMM2 activation via MEKK2 scaffolding. Nat Plants 6:1106–1115
30. Huang Y, Yin C, Liu J et al (2020) A trimeric CrRLK1L-LLG1 complex genetically modulates SUMM2-mediated autoimmunity. Nat Commun 11:4859
31. Liu J, Yang Y, Ortiz-Morea FA et al (2024) The antagonistic role of an E3 ligase pair in regulating plant NLR-mediated autoimmunity and fungal pathogen resistance. Cell Host Microbe 32:1114–1128.e10
32. Yu X, Xu G, Li B et al (2019) The receptor kinases BAK1/SERK4 regulate Ca2+ channel-mediated cellular homeostasis for cell death containment. Curr Biol 29:3778–3790.e8

Chapter 13

Bioinformatics Workflow for Co-Transcriptome Analysis of Plant-Bacterial Interactions

Ying Tang and Kenichi Tsuda

Abstract

Transcriptomic profiling of plant-bacterial interactions provides critical insights into the molecular mechanisms underlying parasitism, commensalism, and mutualism. RNA sequencing (RNA-seq) enables the simultaneous analysis of plant and bacterial transcriptomes during colonization; however, integrated computational workflows specifically tailored for co-transcriptome analysis remain limited. Here, we present a step-by-step bioinformatics pipeline for analyzing co-transcriptome landscapes in plant-bacterial interactions. This workflow includes: (1) quality control and processing of raw RNA-seq data from both plant host and *in-planta* bacterial populations; (2) statistical analyses for differential gene expression; (3) prediction of orthologous bacterial genes and functional annotation of bacterial transcripts using the KEGG database; (4) integration and comparative analysis across multiple bacterial strains; and (5) correlation-based analysis of transcriptional dynamics between plants and bacteria. Designed for researchers with basic familiarity with command-line tools and R programming, this pipeline enables comprehensive analysis of plant-bacterial transcriptional interplay and facilitates hypothesis generation in both pathogenic and symbiotic contexts.

Key words Co-transcriptomics, Plant–bacterial interactions, In-planta bacterial transcriptome, RNA-seq pipeline, Host-pathogen interactions

1 Introduction

The transcriptome represents the complete set of RNA molecules expressed in a cell or tissue under defined conditions, and its analysis allows researchers to uncover gene expression dynamics, identify novel transcripts, and elucidate regulatory networks. RNA sequencing (RNA-seq), a high-throughput and unbiased approach, has revolutionized transcriptomic studies by enabling genome-wide analysis of gene expression across biological systems [1].

Plants coexist with complex bacterial communities, which support plant health or cause disease. These interactions span a continuum from mutualism to parasitism, and the plant immune system plays a central role in shaping microbiota composition and resisting

Tsutomu Kawasaki and Hirofumi Yoshioka (eds.), *Plant Immunity: Methods and Protocols*, Methods in Molecular Biology, vol. 3012,
https://doi.org/10.1007/978-1-0716-5138-4_13,

pathogen invasion [2]. To unravel the molecular dialogues that govern these interactions, co-transcriptomic approaches—simultaneously profiling the transcriptomes of both host and microbe—have emerged as powerful tools. In particular, analysis of the *in-planta* bacterial transcriptome, which represents bacterial gene expression during colonization of plant tissues, is essential for understanding how bacterial behavior adapts to host immunity and environmental conditions. This method reveals bacterial genes that are actively involved in host colonization and influenced by plant immunity.

Since 2018, pioneering studies have demonstrated *in-planta* bacterial RNA-seq as a viable approach to uncovering host-induced bacterial transcriptional reprogramming [2]. This method has enabled the identification of bacterial genes functionally engaged during host-bacterial interactions [3], facilitated integration with proteomic datasets for multi-omics analyses [4], and allowed dissection of co-transcriptome landscapes when combined with plant transcriptome data [5]. While protocols for bacterial RNA isolation from plant tissues are now established [6, 7], standardized and accessible computational pipelines for analyzing co-transcriptome datasets are still lacking. To address this gap, we provide a modular and adaptable bioinformatics workflow for analyzing co-transcriptomes in plant–bacterial systems. Although initially developed for *Arabidopsis thaliana* systems, this pipeline can be readily applied to any plant species with an available genome annotation, thus offering a broadly applicable platform for studying host–bacterial interactions at the transcriptome level.

2 Materials

This section details the datasets, hardware, and software environments required to implement this co-transcriptome analysis pipeline. While default configurations are provided, users may adjust parameters based on specific system resources or requirements.

2.1 Dataset

The pipeline utilizes publicly available RNA-seq data derived from *A. thaliana* and its associated bacterial strains [5]. These data are accessible via the NCBI Gene Expression Omnibus (GEO) under accession number GSE150422 (https://www.ncbi.nlm.nih.gov/geo/query/acc.cgi?acc=GSE150422).

1. Bacterial cDNA libraries were generated using the Ovation Complete Prokaryotic RNA-Seq kit 1–8 (NuGEN). Libraries were sequenced on the Illumina HiSeq3000 system with 1×150 bp single reads. Approximately ten million reads per sample were sequenced.

2. Plant cDNA libraries were generated from poly(A)-enriched RNA using the NEBNext® Ultra™ II Directional RNA Library Prep Kit for Illumina. Libraries were sequenced on the Illumina HiSeq3000 system with 1 × 150 bp single reads. Approximately 30 million reads per sample were sequenced.
3. Reference sequences and annotation files are available from the following sources:

A. thaliana (Araport11, version 202206): The Arabidopsis Information Resource (TAIR, https://www.arabidopsis.org/).

Associated bacteria: At-SPHERE (https://www.at-sphere.com/) [8].

2.2 Hardware

1. A personal computer with at least 8 GB RAM is required for remote access and basic local analysis.
2. Raw sequence data are processed on a remote Unix-based server configured with an Intel Xeon W-2133 processor, 128 GB RAM, and running Ubuntu 18.04.6 LTS. Remote access is established via SSH access from the personal computer (*see* **Note 1**).
3. R-based analyses can be performed either on the server or locally on the personal computer. Prior installation of R and the necessary R packages is required. RStudio is also recommended. Minimum local system specifications for smooth operation include an Intel Core i5 processor, 8 GB RAM, and a 128 GB SSD.

2.3 Software Requirements

The software versions listed below are used to validate this pipeline; later versions may also be compatible, although they have not been explicitly tested.

2.3.1 Software on Remote Linux Server

1. HISAT2 v2.2.1 (https://daehwankimlab.github.io/hisat2/) [9].
2. htseq-count v2.0.3 (https://htseq.readthedocs.io/en/master/htseqcount.html) [10].
3. SAMtools v1.16.1 (https://github.com/samtools/samtools) [11].
4. R version 4.5.0 and RStudio (https://www.r-project.org).

2.3.2 R Packages

1. ClusterProfiler v3.18.1 (https://www.bioconductor.org/packages/release/bioc/html/clusterProfiler) [12].
2. edgeR v4.5.10 (https://bioconductor.org/packages/release/bioc/html/edgeR.html) [13].
3. ggplot2 v3.5.2 (https://ggplot2.tidyverse.org/) [14].

4. ggrepel v0.9.6 (https://cran.r-project.org/web/packages/ggrepel/index) [15].
5. pheamap v1.0.12 (https://cran.r-project.org/web/packages/pheatmap) [16].
6. qvalue v2.39.0 (https://github.com/StoreyLab/qvalue) [17].
7. RColorBrewer v1.1.3 (https://cran.r-project.org/web/packages/RColorBrewer/index.html) [18].

2.3.3 Software on Personal Computer

1. Windows: XShell or PuTTY.
2. macOS: iTerm2 or Royal TSX.

3 Methods

This section outlines the complete pipeline for co-transcriptome analysis, comprising (i) preprocessing of raw RNA-seq data on a remote Linux server and (ii) downstream data analysis in R on a personal computer. To illustrate each step, one *A. thaliana* sample (mock_6h_1) and one bacterial sample (Leaf1) are used as practical examples.

For clarity, the following formatting conventions are applied throughout this section:

- Linux command-line code: dark background
- R code: blue background
- Input files: green
- Output files: orange

All steps are executed on a remote Linux server, while R analyses can be carried out locally in the R environment (*see* **Note 2**).

3.1 Plant Transcriptomics

This section describes the processing steps for plant RNA-seq data, including alignment, quantification, and downstream analysis.

3.1.1 Data Download and Integrity Verification

The RNA-seq data for the sample mock_6h_1 consists of two single-end read files, which can be downloaded from the EMBL-EBI database using the following commands:

```
$ wget ftp.sra.ebi.ac.uk/vol1/fastq/SRR117/080/SRR11780380/SRR11780380.fastq.gz
```

```
$ wget ftp.sra.ebi.ac.uk/vol1/fastq/SRR117/081/SRR11780381/SRR11780381.fastq.gz
```

```
d58e05ef33707223b34c54a77366a5a3  SRR11780380.fastq.gz
dd7273ffbf7edcab02d5fbfef1b6ff07  SRR11780381.fastq.gz
```

Fig. 1 MD5 value of samples

```
arabi.1.ht2  arabi.3.ht2  arabi.5.ht2  arabi.7.ht2
arabi.2.ht2  arabi.4.ht2  arabi.6.ht2  arabi.8.ht2
```

Fig. 2 Outputs of the indexing command

To verify download integrity, calculate the MD5 checksums:

```
$ md5sum SRR11780380.fastq.gz
```

```
$ md5sum SRR11780381.fastq.gz
```

Compare the resulting checksums with those provided in the repository. Proceed only if they match, confirming that the files were downloaded completely and without errors (*see* Fig. 1).

3.1.2 Indexing and Read Mapping

First, build the HISAT2 index for the *A. thaliana* reference genome (Athaliana_447_TAIR10.fa), using the prefix arabi:

```
$ hisat2-build -p 8 Athaliana_447_TAIR10.fa arabi
```

This command generates eight index files with the prefix arabi (*see* Fig. 2).

Next, map the reads to the reference genome using HISAT2. The output is streamed directly into SAMtools for BAM conversion and sorting:

```
$ hisat2 -p 20 --rna-strandness R arabi -U SRR11780380.fastq.gz,SRR11780381.fastq.gz |
```

```
samtools view -@ 20 -bh | samtools sort -@ 20 > mock_1.bam
```

To evaluate mapping quality, use the following command:

```
$ samtools flagstat mock_1.bam
```

In this example, 12,255,559 reads were successfully mapped, corresponding to a mapping rate of 99.2% (*see* Fig. 3).

```
12255559 + 0 in total (QC-passed reads + QC-failed reads)
236388 + 0 secondary
0 + 0 supplementary
0 + 0 duplicates
12157051 + 0 mapped (99.20% : N/A)
0 + 0 paired in sequencing
0 + 0 read1
0 + 0 read2
0 + 0 properly paired (N/A : N/A)
0 + 0 with itself and mate mapped
0 + 0 singletons (N/A : N/A)
0 + 0 with mate mapped to a different chr
0 + 0 with mate mapped to a different chr (mapQ>=5)
```

Fig. 3 Mapping quality of the example mock_1.bam

AT1G01010	122
AT1G01020	94
AT1G01030	33
AT1G01040	544
AT1G01046	0
AT1G01050	391
AT1G01060	12
AT1G01070	46
AT1G01080	941

Fig. 4 An excerpt of the resulting count table

3.1.3 Gene Expression Quantification

To quantify gene expression, use HTSeq-count on the aligned BAM file, along with the Araport11 gene annotation file (Araport11_GTF_genes_transposons.current.gtf):

```
$htseq-count -s reverse -q -t exon mock_1.bam

Araport11_GTF_genes_transposons.current.gtf > mock_1.txt
```

This command generates a gene-level count table (mock_1.txt) suitable for downstream differential expression analysis. An excerpt of the resulting count table is shown in Fig. 4.

3.1.4 Detection of Differentially Expressed Genes (DEGs)

After generating count tables for all samples, combine them into a single input file. For example, 3.1.4_input_plantcount.txt includes gene expression counts for *A. thaliana* control and treatment groups (e.g., mock and Pst DC3000). A portion of this count table is shown in Fig. 5.

	mock_6h_1	mock_6h_2	mock_6h_3	D36E_6h_1	D36E_6h_2	D36E_6h_3
AT1G01010	122	218	156	2111	1758	1561
AT1G01020	94	193	130	188	131	103
AT1G01030	33	54	22	129	92	75
AT1G01040	544	963	945	1350	1084	818
AT1G01046	0	0	1	0	0	0
AT1G01050	391	1067	982	246	288	211
AT1G01060	12	3	4	18	21	11
AT1G01070	46	42	33	115	80	95
AT1G01080	941	2055	1918	80	144	129

Fig. 5 A portion of the input, 3.1.4_input_plantcount.txt

To identify DEGs, the DESeq2 package in R is used as follows:

```
library(DESeq2)
data <- read.table("3.1.4_input_plantcount.txt", header = T,row.names = 1,check.names = F)
condition <- factor(c(rep("control",3),rep("treat",3)), levels = c("control", "treat"))
colData <- data.frame(row.names=colnames(data), condition)
dds <- DESeqDataSetFromMatrix(data, colData, design= ~ condition)
dds <- DESeq(dds)
res=results(dds)
summary(res)
resSig <- subset(res,padj<0.05)
up <- subset(resSig, log2FoldChange > 1)
down <- subset(resSig, log2FoldChange < -1)
write.table(resSig, "3.1.4_output_DEG.txt", sep="\t", quote=F)
```

baseMean	log2FoldChange	lfcSE	stat	pvalue	padj	
AT1G01010	960.6046512	3.44225222	0.16459621	20.9133138	4.05E-97	3.18E-95
AT1G01030	65.78148949	1.37637128	0.37099492	3.70994644	0.0002073	0.00043554
AT1G01040	917.4975702	0.39325159	0.12718424	3.09198366	0.00198824	0.0037252
AT1G01050	499.0824077	-1.6384902	0.20935385	-7.8264153	5.02E-15	2.24E-14
AT1G01070	69.03518192	1.15788152	0.37709518	3.07052856	0.0021368	0.00398687
AT1G01080	837.4803416	-3.7173192	0.272207	-13.656222	1.85E-42	2.93E-41
AT1G01090	1718.616599	-1.0008199	0.15813179	-6.3290242	2.47E-10	8.38E-10
AT1G01120	2063.179845	-1.1461635	0.25019274	-4.581122	4.62E-06	1.14E-05
AT1G01140	879.055502	-2.1967726	0.1357389	-16.18381	6.56E-59	1.93E-57

Fig. 6 A partial view of the output, 3.1.4_output_DEG.txt

AT1G01010	regulation of DNA-templated transcription	GO:0006355	P
AT1G01010	defense response to other organism	GO:0098542	P
AT1G01010	response to oxidative stress	GO:0006979	P
AT1G01010	regulation of DNA-templated transcription	GO:0006355	P
AT1G01010	response to abscisic acid	GO:0009737	P
AT1G01010	response to lipid	GO:0033993	P
AT1G01010	oxoacid metabolic process	GO:0043436	P
AT1G01010	regulation of DNA-templated transcription	GO:0006355	P
AT1G01010	defense response to other organism	GO:0098542	P
AT1G01010	response to inorganic substance	GO:0010035	P
AT1G01010	regulation of DNA-templated transcription	GO:0006355	P
AT1G01010	defense response to other organism	GO:0098542	P

Fig. 7 A portion of the input, 3.1.5_input_GOannotation.txt

This output file 3.1.4_output_DEG.txt contains a list of DEGs, including $\log_2$ fold changes and adjusted *p*-values. A partial view of the output table is provided in Fig. 6.

3.1.5 Gene Ontology (GO) Enrichment Analysis

Up- and down-regulated gene sets identified in the previous step can be used for GO enrichment analysis. The input file 3.1.5_input_GOannotation.txt (a portion is shown as Fig. 7) contains GO annotations (GO terms, ontology categories, gene IDs).

The following R code performs GO enrichment analysis for upregulated genes using the clusterProfiler, DOSE, and GO.db packages:

```
library(DOSE)
library(clusterProfiler)
library(GO.db)
library(ggplot2)
ATH_GOTERM <- read.table("3.1.5_input_GOannotation.txt", sep = "\t", quote="")
colnames(ATH_GOTERM) <- c("proID", "Source", "GOTerm", "Ont")


ATH_GOTERM_BP <- ATH_GOTERM[ATH_GOTERM$Ont == "P",]
gomap_BP <- data.frame(GO=ATH_GOTERM_BP$GOTerm,
gene=ATH_GOTERM_BP$proID)
goname_BP <- AnnotationDbi::select(x=GO.db, keys = gomap_BP$GO, keytype = "GOID",
columns = "TERM")
ego_BP <- enricher(rownames(up), TERM2GENE = gomap_BP, TERM2NAME =
goname_BP)
ego_BP@result$Ont <- c("BP")
ATH_GOTERM_MF <- ATH_GOTERM[ATH_GOTERM$Ont == "F",]
gomap_MF <- data.frame(GO=ATH_GOTERM_MF$GOTerm,
gene=ATH_GOTERM_MF$proID)
goname_MF <- AnnotationDbi::select(x=GO.db, keys = gomap_MF$GO, keytype = "GOID",
columns = "TERM")
ego_MF <- enricher(rownames(up), TERM2GENE = gomap_MF, TERM2NAME =
```

```
goname_MF)
ego_MF@result$Ont <- c("MF")
ATH_GOTERM_CC <- ATH_GOTERM[ATH_GOTERM$Ont == "C",]
gomap_CC <- data.frame(GO=ATH_GOTERM_CC$GOTerm,
gene=ATH_GOTERM_CC$proID)
goname_CC <- AnnotationDbi::select(x=GO.db, keys = gomap_CC$GO, keytype = "GOID",
columns = "TERM")
ego_CC <- enricher(rownames(up), TERM2GENE = gomap_CC, TERM2NAME =
goname_CC)
ego_CC@result$Ont <- c("CC")


ego_PriTerML <- rbind(ego_BP@result, ego_MF@result, ego_CC@result)


p <- dotplot(merge_result(list(inter=ego_PriTerML)), x="GeneRatio", size="count", split="Ont",
font.size=12, label_format=50) + facet_grid(Ont~., scale="free") + theme(axis.text.y =
element_text(face = "bold"))
#plot <- merge_result(list(PriTerML=ego_PriTerML, pri=ego_pri, ter=ego_ter, ML=ego_ML,
VennPri=ego_VennPri, VennTer=ego_VennTer, VennML=ego_VennML,
VennPriTer=ego_VennPriTer, VennPriML=ego_VennPriML, VennTerML=ego_VennTerML,
VennPriTerML=ego_VennPriTerML))
ggsave("3.1.5_output_upGOenrichemnt.png", p, height=8, width = 8)
```

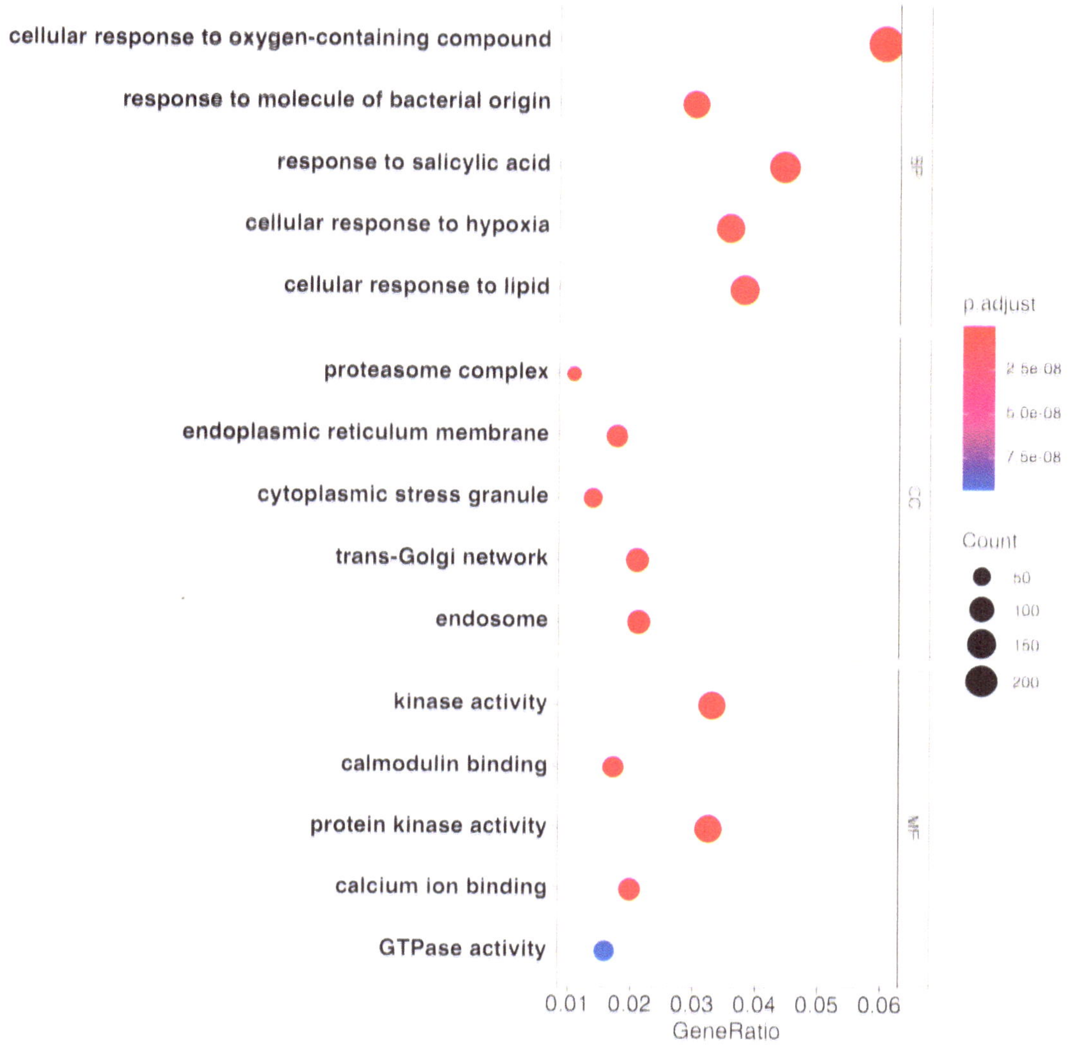

Fig. 8 3.1.5_output_upGOenrichment.png, A dot plot for GO enrichment analysis visualization

This script produces 3.1.5_output_upGOenrichment.png, a dot plot summarizing enriched GO terms among upregulated genes, categorized by ontology (*see* Fig. 8).

3.2 In-Planta Bacterial Transcriptomics

The analysis pipeline for bacterial transcriptomes follows the same general workflow as that for plant transcriptomes, with the primary distinction being the use of bacterial reference genomes and annotations. For bacterial samples, genome assemblies and annotations from the At-SPHERE database were used. In this section, the bacterial sample Leaf1 is used to illustrate the analysis process.

3.2.1 Data Download to Gene Count

Download and validate the RNA-seq data for Leaf1 using the following commands:

```
$ wget ftp.sra.ebi.ac.uk/vol1/fastq/SRR117/092/SRR11780292/SRR11780292.fastq.gz
```

```
$ wget ftp.sra.ebi.ac.uk/vol1/fastq/SRR117/092/SRR11780292/SRR11780293.fastq.gz
```

Verify download integrity using MD5 checksum validation:

```
$ md5sum SRR11780292.fastq.gz
```

```
$ md5sum SRR11780293.fastq.gz
```

Build the HISAT2 index using the bacterial reference genome (Leaf1.fna):

```
$ hisat2-build -p 8 Leaf1.fna Leaf1
```

Align the reads to the bacterial genome and convert the output to a sorted BAM file:

```
$ hisat2 -p 20 --rna-strandness R Leaf1 -U SRR11780292.fastq.gz, SRR11780293.fastq.gz |
```

```
samtools view -@ 20 -bh | samtools sort -@ 20 > Leaf1_1.bam
```

Use HTSeq-count to quantify gene expression based on the aligned reads and the bacterial annotation file (Leaf1.gff):

```
$htseq-count -s reverse -q -t exon Leaf1_1.bam Leaf1.gff > Leaf1_1.txt
```

This results in Leaf1_1.txt, a gene-level count table representing the expression of bacterial genes *in planta*. A portion of the resulting table is shown in Fig. 9.

3.2.2 Functional Annotation Enrichment

To enable cross-strain comparisons of gene expression among bacterial strains, orthologous groups (OGs) and KEGG orthologs (KOs) must first be identified. This involves two key steps:

- Orthology inference via OrthoFinder using predicted protein sequences.
- Functional annotation via KEGG-based KO assignment using KofamScan.

cds0	0
cds1	3
cds10	0
cds100	189
cds1000	1
cds1001	138
cds1002	0
cds1003	0
cds1004	124
cds1005	409

Fig. 9 A portion of the resulting table, Leaf1_1.txt

Orthogroup	Leaf1	Leaf130	Leaf155	Leaf176	Leaf177	Leaf187	Leaf404	Root935	Soil763
OG0000000		Leaf130|1027, Leaf130|1048, Leaf130|1073, Leaf130|1103, Leaf130|1116, Leaf130|1145, Leaf130|1471, Leaf130|1782, Leaf130|223, Leaf130|2628, Leaf130|364, Leaf130|435, Leaf130|4							
OG0000001					Leaf177|2682, Leaf177|2683, Leaf177|2692, Leaf177|3204, Leaf177|3302, Leaf177|3303, Leaf177|3752, Leaf177|3753, Leaf177|3754, Leaf177|37				
OG0000002	Leaf1|1372, Leaf1|1702, Leaf1|2039, Leaf1|2040, Leaf1|2102, Leaf1|2103, Leaf1|2206, Leaf1|2222, Leaf1|2328, Leaf1|262, Leaf1|2844								
OG0000003				Leaf176|1124, Leaf176|1125, Leaf176|1126, Leaf176|1127, Leaf176|1128, Leaf176|1129, Leaf176|1130, Leaf176|1131, Leaf176|1132					
OG0000004			Leaf155|1699, Leaf155|1745, Leaf155|1829, Leaf155|1880, Leaf155|2038, Leaf155|2386, Leaf155|3011, Leaf155|3114						
OG0000005					Leaf177|1321, Leaf177|1322, Leaf177|2798, Leaf177|3109, Leaf177|3814, Leaf177|4787, Leaf177|5542, Leaf177|5803				
OG0000006	Leaf1|533		Leaf155|3830		Leaf177|2101, Leaf177|3518, Leaf177|5232, Leaf177				Soil763|625
OG0000007	Leaf1|1582		Leaf155|1682, Leaf155|17		Leaf177|1608, Leaf177|1621, Leaf177|1622				Soil763|1225
OG0000008				Leaf176|1107, Leaf176|1117, Leaf176|3194, Leaf176|3197, Leaf176|3205, Leaf176|3209, Leaf176|4120					
OG0000009					Leaf177|1940, Leaf177|2850, Leaf177|3585, Leaf177|4540, Leaf177|4824, Leaf177|4985, Leaf177|909				
OG0000010							Leaf404|1122, Leaf404|2339, Leaf404|3131, Leaf404|3465, Leaf404|3466, Leaf404|3467, Leaf404|3470		
OG0000011	Leaf1|1262		Leaf155|1424						Soil763|1487, Soil763|2152, Soil763|2153, Soil763|3156
OG0000012			Leaf155|35		Leaf177|2487, Leaf177|968				Soil763|106, Soil763|1088, Soil763|3080
OG0000013			Leaf155|2834, Leaf155|2910, Leaf155|3858, Leaf155|387, Leaf155|3925, Leaf155|4388						
OG0000014				Leaf176|1914, Leaf176|1915, Leaf176|1916, Leaf176|1919, Leaf176|1935, Leaf176|2415					
OG0000015					Leaf177|1343, Leaf177|1812, Leaf177|2151, Leaf177|2152, Leaf177|2593, Leaf177|4135				
OG0000016									Soil763|1162, Soil763|1163, Soil763|1164, Soil763|1165, Soil763|1166, Soil763|1167

Fig. 10 A portion of OG annotation, the result of the tool orthofinder

Run the following commands:

```
$ orthofinder -d -f strains.faa
```

```
$ kofamscan -o kofam.out --format mapper strains.faa
```

These tools generate consistent functional annotations (OG and KO) across bacterial strains, enabling the integration of gene expression data at the level of orthologous groups and metabolic pathways (via KO terms) (*see* Figs. 10 and 11).

3.2.3 Obtaining the DEGs (KO Terms)

To facilitate functional enrichment analysis, KO terms corresponding to DEGs are extracted from a summary file, such as 3.2.3_input_Leaf1summary.txt. This file includes expression values *in planta* and in vitro, $\log_2$ fold changes, DEG classification (e.g., up- or down-regulated), and KO annotations (*see* Fig. 12).

<table>
<tr><td>OG0000084</td><td>OG0000084|K10440|rbsC; ribose transport system permease protein</td></tr>
<tr><td>OG0000054</td><td>OG0000054|K02624|pcaR; IclR family transcriptional regulator, pca regulon regulatory protein</td></tr>
<tr><td>OG0000269</td><td>OG0000269|K13979|yahK; uncharacterized zinc-type alcohol dehydrogenase-like protein [EC:1.-.-.-]</td></tr>
<tr><td>OG0000022</td><td>OG0000022|K13653|AraC family transcriptional regulator</td></tr>
<tr><td>OG0001477</td><td>OG0001477||</td></tr>
<tr><td>OG0000689</td><td>OG0000689|K05782|benE; benzoate membrane transport protein</td></tr>
<tr><td>OG0002877</td><td>OG0002877|K19746|dauA; D-arginine dehydrogenase [EC:1.4.99.6]</td></tr>
<tr><td>OG0000186</td><td>OG0000186|K09022|ridA; 2-iminobutanoate/2-iminopropanoate deaminase [EC:3.5.99.10]</td></tr>
<tr><td>OG0003738</td><td>OG0003738|K01733|thrC; threonine synthase [EC:4.2.3.1]</td></tr>
<tr><td>OG0000237</td><td>OG0000237|K00600|glyA; glycine hydroxymethyltransferase [EC:2.1.2.1]</td></tr>
<tr><td>OG0002110</td><td>OG0002110|K00311|ETFDH; electron-transferring-flavoprotein dehydrogenase [EC:1.5.5.1]</td></tr>
<tr><td>OG0000994</td><td>OG0000994|K03521|fixA; electron transfer flavoprotein beta subunit</td></tr>
<tr><td>OG0000690</td><td>OG0000690|K03522|fixB; electron transfer flavoprotein alpha subunit</td></tr>
<tr><td>OG0002878</td><td>OG0002878|K02030|ABC.PA.S; polar amino acid transport system substrate-binding protein</td></tr>
<tr><td>OG0003219</td><td>OG0003219||</td></tr>
</table>

Fig. 11 A portion of KO annotation, the result of the tool kofamscan

	L1_Col	L1_TSB	L1_FC	DEG	OG	KO.id	KO.description
cds100	10.2060777	9.60259873	1.19953431	up	OG0001554	K13788	pta; phosphate acetyltransferase [EC:2.3.1.8]
cds1000	4.21018307	5.46611822	-0.4641249	nonDEG	OG0000015	K02031	ABC.PE.A; peptide/nickel transport system ATP-binding protein
cds1001	9.88964167	11.0215995	-0.5323686	nonDEG	OG0000104	K16568	exoZ; exopolysaccharide production protein ExoZ
cds1003	5.56071849	6.05873011	-1.7474905	nonDEG	OG0000109		
cds1004	9.29732274	7.40027192	2.51679017	up			
cds1005	11.4679969	10.7914023	1.28207858	up	OG0000374	K01414	prlC; oligopeptidase A [EC:3.4.24.70]
cds1006	8.36145748	7.81783292	1.182902	nonDEG	OG0000705	K03544	clpX; ATP-dependent Clp protease ATP-binding subunit ClpX
cds1007	11.1978842	10.3374165	1.47874947	up	OG0000451	K01358	clpP; ATP-dependent Clp protease, protease subunit [EC:3.4.21.92]
cds1008	4.08093222	6.30372495	-1.7711167	nonDEG	OG0000451	K01358	clpP; ATP-dependent Clp protease, protease subunit [EC:3.4.21.92]
cds1009	8.7854846	10.7322994	-1.3446739	down			
cds1012	12.3513075	9.7210281	3.25304865	up	OG0000373		
cds1013	5.77709093	6.97240818	-0.5468268	nonDEG	OG0000477	K04047	dps; starvation-inducible DNA-binding protein
cds1014	7.61405718	7.98854976	0.26446838	nonDEG	OG0000176		
cds1015	6.96365924	8.19701037	-0.4234925	nonDEG	OG0000461	K10563	mutM; formamidopyrimidine-DNA glycosylase [EC:3.2.2.23 4.2.99.18]
cds1016	8.79975869	8.30984758	1.08230791	nonDEG	OG0001887	K01808	rpiB; ribose 5-phosphate isomerase B [EC:5.3.1.6]

Fig. 12 A portion of the input, 3.2.3_input_Leaf1summary.txt

The following R script separates KO terms for upregulated and downregulated genes:

```
data <- read.delim("3.2.3_input_Leaf1summary.txt", header = T, row.names = 1)

up <- data[data[, 4]=="up"   ,]

down <-data[data[, 4]=="down" ,]

indata_up <- na.omit(up[, 6])

indata_up <- as.data.frame(indata_up)

indata_down <- na.omit(down[, 6])

indata_down <- as.data.frame(indata_down)
```

3.2.4 KO Enrichment Analysis for DEGs

To identify biologically meaningful patterns among DEGs, enrichment analysis of KO terms is performed using a custom function ko.enrich.strain(). This function applies a hypergeometric test followed by false discovery rate (FDR) correction to determine significantly enriched KO terms.

The analysis focuses on KO terms within functional category level-3, meeting the following criteria:

- Annotated by at least 5 DEGs
- Fold enrichment >2
- Adjusted *p*-value <0.01

As an example, we demonstrate this analysis using Leaf1. The KO annotation database for this strain (3.2.4_input_Leaf1_KOdatabase.txt) is formatted with KO terms, associated functional categories, and gene IDs (*see* Fig. 13).

The custom enrichment function is defined as follows, full code available at: https://github.com/YTang906/co-transcriptome (*see* **Note 3**).

```
ko.enrich.strain <- function(test_ko_list, strain){ ...}
```

Run the function on Leaf1:

```
hypr <- ko.enrich.strain(indata_up, "Leaf1")
hypr[, 6] <- "up"
colnames(hypr)[6] <- "updown"
hypr2 <-ko.enrich.strain(indata_down, "Leaf1")
hypr2[, 6] <- "down"
colnames(hypr2)[6] <- "updown"

out <- rbind(hypr, hypr2)
write.table(out, file="output/dir/3.2.4_output_Leaf1_category3.txt", sep = ""), row.names=F,
sep="\t", quote=F)
```

KO	category-1	category-2	category-3	category-4	N of genes
K00003	Metabolism	Cysteine and methionine metabolism	Cysteine and methionine metabolism	hom; homoserine dehydrogenase [EC:1.1.1.3]	1
K00003	Metabolism	Glycine, serine and threonine metabolism	Glycine, serine and threonine metabolism	hom; homoserine dehydrogenase [EC:1.1.1.3]	1
K00003	Metabolism	Lysine biosynthesis	Lysine biosynthesis	hom; homoserine dehydrogenase [EC:1.1.1.3]	1
K00009	Metabolism	Fructose and mannose metabolism	Fructose and mannose metabolism	mtlD; mannitol-1-phosphate 5-dehydrogenase [EC:1.1.1.17]	4
K00010	Metabolism	Inositol phosphate metabolism	Inositol phosphate metabolism	iolG; myo-inositol 2-dehydrogenase / D-chiro-inositol 1-dehydrogenase [EC:1.1.1.18	7
K00010	Metabolism	Streptomycin biosynthesis	Streptomycin biosynthesis	iolG; myo-inositol 2-dehydrogenase / D-chiro-inositol 1-dehydrogenase [EC:1.1.1.18	7
K00012	Metabolism	Amino sugar and nucleotide sugar metabolism	Amino sugar and nucleotide sugar metabolism	UGDH, ugd; UDPglucose 6-dehydrogenase [EC:1.1.1.22]	1
K00012	Metabolism	Ascorbate and aldarate metabolism	Ascorbate and aldarate metabolism	UGDH, ugd; UDPglucose 6-dehydrogenase [EC:1.1.1.22]	1
K00012	Metabolism	Pentose and glucuronate interconversions	Pentose and glucuronate interconversions	UGDH, ugd; UDPglucose 6-dehydrogenase [EC:1.1.1.22]	1
K00013	Metabolism	Histidine metabolism	Histidine metabolism	hisD; histidinol dehydrogenase [EC:1.1.1.23]	2
K00014	Metabolism	Phenylalanine, tyrosine and tryptophan biosynthesis	Phenylalanine, tyrosine and tryptophan biosynthesis	aroE; shikimate dehydrogenase [EC:1.1.1.25]	1
K00020	Metabolism	Valine, leucine and isoleucine degradation	Valine, leucine and isoleucine degradation	mmsB, HIBADH; 3-hydroxyisobutyrate dehydrogenase [EC:1.1.1.31]	1
K00024	Metabolism	Citrate cycle (TCA cycle)	Citrate cycle (TCA cycle)	mdh; malate dehydrogenase [EC:1.1.1.37]	2
K00024	Metabolism	Glyoxylate and dicarboxylate metabolism	Glyoxylate and dicarboxylate metabolism	mdh; malate dehydrogenase [EC:1.1.1.37]	2
K00024	Metabolism	Pyruvate metabolism	Pyruvate metabolism	mdh; malate dehydrogenase [EC:1.1.1.37]	2

Fig. 13 A portion of the input, 3.2.4_input_Leaf1_KOdatabase.txt

<table>
<tr><th>KO</th><th>pval</th><th>FoldEnrichment</th><th>genes</th><th>size</th><th>updown</th></tr>
<tr><td>ABC transporters</td><td>3.79E-06</td><td>3.347091694</td><td>K17317---gtsC, glcG; glucose/mannose transport system permease protein|K10117---ms</td><td>15</td><td>up</td></tr>
<tr><td>DNA repair and recombination proteins [BR:ko03400]</td><td>0</td><td>6.917322835</td><td>K03111---ssb; single-strand DNA-binding protein|K14161---imuB; protein ImuB|K02337</td><td>17</td><td>up</td></tr>
<tr><td>Enzymes with EC numbers</td><td>4.16E-11</td><td>4.693897638</td><td>K03207---wcaH; colanic acid biosynthesis protein WcaH [EC:3.6.1.-]|K18455---mca; myco</td><td>19</td><td>up</td></tr>
<tr><td>Exosome [BR:ko04147]</td><td>0</td><td>6.917322835</td><td>K01897---ACSL, fadD; long-chain acyl-CoA synthetase [EC:6.2.1.3]|K00088---IMPDH, gu</td><td>13</td><td>up</td></tr>
<tr><td>Fructose and mannose metabolism</td><td>0</td><td>20.7519685</td><td>K01805---xylA; xylose isomerase [EC:5.3.1.5]|K01803---TPI, tpiA; triosephosphate isomer</td><td>6</td><td>up</td></tr>
<tr><td>Function unknown</td><td>0</td><td>41.50393701</td><td>K09118---K09118; uncharacterized protein|K09932---K09932; uncharacterized protein|K0</td><td>6</td><td>up</td></tr>
<tr><td>General function prediction only</td><td>0</td><td>9.684251969</td><td>K06876---K06876; deoxyribodipyrimidine photolyase-related protein|K09858---K09858;</td><td>7</td><td>up</td></tr>
<tr><td>Glutathione metabolism</td><td>0</td><td>6.917322835</td><td>K00031---IDH1, IDH2, icd; isocitrate dehydrogenase [EC:1.1.1.42]|K00797---speE, SRM; sp</td><td>6</td><td>up</td></tr>
<tr><td>Homologous recombination</td><td>0</td><td>34.58661417</td><td>K03111---ssb; single-strand DNA-binding protein|K02337---dnaE; DNA polymerase III s</td><td>5</td><td>up</td></tr>
<tr><td>Lipid biosynthesis proteins [BR:ko01004]</td><td>0.00696064</td><td>3.144237652</td><td>K01897---ACSL, fadD; long-chain acyl-CoA synthetase [EC:6.2.1.3]|K00059---fabG; 3-ox</td><td>5</td><td>up</td></tr>
<tr><td>Mismatch repair</td><td>0.001716</td><td>3.84295713</td><td>K03111---ssb; single-strand DNA-binding protein|K02337---dnaE; DNA polymerase III s</td><td>5</td><td>up</td></tr>
<tr><td>Pentose phosphate pathway</td><td>0</td><td>48.42125984</td><td>K00036---G6PD, zwf; glucose-6-phosphate 1-dehydrogenase [EC:1.1.1.49 1.1.1.363]|K00</td><td>7</td><td>up</td></tr>
<tr><td>Purine metabolism</td><td>0</td><td>8.646653543</td><td>K01951---guaA, GMPS; GMP synthase (glutamine-hydrolysing) [EC:6.3.5.2]|K00088---IM</td><td>5</td><td>up</td></tr>
<tr><td>Quorum sensing</td><td>0</td><td>29.97506562</td><td>K01897---ACSL, fadD; long-chain acyl-CoA synthetase [EC:6.2.1.3]|K01997---livH; branc</td><td>13</td><td>up</td></tr>
</table>

Fig. 14 A partial of the output, 3.2.4_output_Leaf1_category3.txt

The output file 3.2.4_output_Leaf1_category3.txt contains a list of significantly enriched KO terms, including enrichment statistics and classification (up or down). A partial view of the output is shown in Fig. 14.

3.2.5 KO Enrichment Analysis for Various Strains

The KO enrichment analysis function introduced in Subheading 3.2.4 can be extended to multiple bacterial strains in a batch-processing framework. This enables systematic identification of enriched KOs across bacterial strains.

The following R script performs KO enrichment analysis for a list of strains by iterating through their corresponding DEG summary files:

```
dir_data <- "data/input/3.2.5/"
out_dir <- "data/output/3.2.5_"
strain <- c("Leaf1","Leaf130", "Leaf155","Leaf177", "Leaf187", "Leaf176", "Leaf404",
"Root935", "Soil763", "Pto", "D36E", "MM")

for (i in c(1:length(strain))){
    data <- read.delim(paste(dir_data, "3.2.5_input_", strain[i], "_summary.txt", sep = "" ),
header = T, row.names = 1)
    up <- data[data[, 4]=="up"  ,]
    down <-data[data[, 4]=="down" ,]
    indata_up <- na.omit(up[, 6])
    indata_up <- as.data.frame(indata_up)
    indata_down <- na.omit(down[, 6])
    indata_down <- as.data.frame(indata_down)

    # KO enrichment analysis
    hypr <- ko.enrich.strain(indata_up, strain[i])
    hypr[, 6] <- "up"
    colnames(hypr)[6] <- "updown"
    hypr2 <-ko.enrich.strain(indata_down, strain[i])
    hypr2[, 6] <- "down"
    colnames(hypr2)[6] <- "updown"
    # output results
>   out <- rbind(hypr, hypr2)
    write.table(out, file=paste(out_dir, "3.2.5_output_", strain[i],"_category3.txt", sep = ""),
row.names=F,  sep="\t", quote=F)
}
```

This pipeline systematically generates KO enrichment summaries for each strain, with output files named as 3.2.5_output_<strain>_category3.txt, enabling downstream comparative analysis of functionally enriched gene sets across diverse bacterial taxa.

3.2.6 Visualization of KO Term Expression in Differentially Enriched KOs (DEKOs)

To facilitate the comparative interpretation of functional shifts across bacterial strains and treatment conditions, the expression of DEGs associated with DEKOs can be visualized using ggplot2. The function ko.plot() automates this process by:

1. Extracting fold change and DEG classification across bacterial strains
2. Overlaying phylogenetic metadata to assign phylum-level taxonomy
3. Visualizing gene expression using boxplots with color-coded points for DEG classification

Inputs

- A KO master annotation file: 3.2.6_input_KO_master_database.txt (*see* Fig. 15)
- DEG summary files for each strain (used in Subheading 3.2.3), which include $\log_2$ fold change and DEG classification
- A strain taxonomy file: 3.2.6_input_commensal_phylogeny.txt
- A KO pathway annotation list: 3.2.6_input_ko_pathway_annotation_list_for_plots.txt

The following R script generates annotated boxplots summarizing DEG fold changes across strains for the target KO terms:

K00001	Metabolism	Glycolysis / Gluconeogenesis	Glycolysis / Gluconeogenesis	E1.1.1.1, adh; alcohol dehydrogenase [EC:1.1.1.1]
K00001	Metabolism	Fatty acid degradation	Fatty acid degradation	E1.1.1.1, adh; alcohol dehydrogenase [EC:1.1.1.1]
K00001	Metabolism	Tyrosine metabolism	Tyrosine metabolism	E1.1.1.1, adh; alcohol dehydrogenase [EC:1.1.1.1]
K00001	Metabolism	Retinol metabolism	Retinol metabolism	E1.1.1.1, adh; alcohol dehydrogenase [EC:1.1.1.1]
K00001	Metabolism	Chloroalkane and chloroalkene degradation	Chloroalkane and chloroalkene degradation	E1.1.1.1, adh; alcohol dehydrogenase [EC:1.1.1.1]
K00001	Metabolism	Drug metabolism - cytochrome P450	Drug metabolism - cytochrome P450	E1.1.1.1, adh; alcohol dehydrogenase [EC:1.1.1.1]
K00001	Metabolism	Metabolism of xenobiotics by cytochrome P450	Metabolism of xenobiotics by cytochrome P450	E1.1.1.1, adh; alcohol dehydrogenase [EC:1.1.1.1]
K00001	Metabolism	Naphthalene degradation	Naphthalene degradation	E1.1.1.1, adh; alcohol dehydrogenase [EC:1.1.1.1]
K00002	Metabolism	Glycolysis / Gluconeogenesis	Glycolysis / Gluconeogenesis	AKR1A1, adh; alcohol dehydrogenase (NADP+) [EC:1.1.1.2]
K00002	Metabolism	Pentose and glucuronate interconversions	Pentose and glucuronate interconversions	AKR1A1, adh; alcohol dehydrogenase (NADP+) [EC:1.1.1.2]
K00002	Metabolism	Glycerolipid metabolism	Glycerolipid metabolism	AKR1A1, adh; alcohol dehydrogenase (NADP+) [EC:1.1.1.2]
K00002	Metabolism	Caprolactam degradation	Caprolactam degradation	AKR1A1, adh; alcohol dehydrogenase (NADP+) [EC:1.1.1.2]
K00002	Brite Hierarchies	Exosome [BR:ko04147]	Exosome [BR:ko04147]	AKR1A1, adh; alcohol dehydrogenase (NADP+) [EC:1.1.1.2]
K00003	Metabolism	Cysteine and methionine metabolism	Cysteine and methionine metabolism	hom; homoserine dehydrogenase [EC:1.1.1.3]
K00003	Metabolism	Glycine, serine and threonine metabolism	Glycine, serine and threonine metabolism	hom; homoserine dehydrogenase [EC:1.1.1.3]
K00003	Metabolism	Lysine biosynthesis	Lysine biosynthesis	hom; homoserine dehydrogenase [EC:1.1.1.3]
K00004	Metabolism	Butanoate metabolism	Butanoate metabolism	BDH, butB; (R,R)-butanediol dehydrogenase / meso-butanediol deh
K00005	Metabolism	Propanoate metabolism	Propanoate metabolism	gldA; glycerol dehydrogenase [EC:1.1.1.6]
K00005	Metabolism	Glycerolipid metabolism	Glycerolipid metabolism	gldA; glycerol dehydrogenase [EC:1.1.1.6]
K00006	Metabolism	Glycerophospholipid metabolism	Glycerophospholipid metabolism	GPD1; glycerol-3-phosphate dehydrogenase (NAD+) [EC:1.1.1.8]
K00006	Environmental Info	MAPK signaling pathway - yeast	MAPK signaling pathway - yeast	GPD1; glycerol-3-phosphate dehydrogenase (NAD+) [EC:1.1.1.8]

Fig. 15 A partial of the input, 3.2.6_input_KO_master_database.txt

```
library(ggplot2)


#input is a vector of KOs
ko_plot <- function(target){
  target <- unique(target)
  dir_data <- "data/input/3.2.6/"
  strain2 <- c("Leaf1","Soil763","Leaf130", "Leaf155","Leaf177", "Leaf187", "Leaf176", "Leaf404", "Root935",    "D36E", "MM", "Pto"  )
  strain <- c("D36E", "Leaf1", "Leaf130", "Leaf155", "Leaf176", "Leaf177", "Leaf187", "Leaf404", "MM", "Pto", "Root935", "Soil763")

  # Step1: upload needed information of genes in each strain
  ko_database <- read.delim(file = paste(dir_data, "3.2.6_input_KO_master_database.txt", sep = ""),   header = F)

  out_fc <- data.frame(matrix(vector(), 0, length(strain),
                        dimnames = list(c(), strain)),
                        stringsAsFactors = F) # Fold change of genes in each strain
  out_deg <- data.frame(matrix(vector(), 0, length(strain),
                              dimnames = list(c(), strain)),
                        stringsAsFactors = F) # Classification of genes in each strain

  for (i in c(1:length(strain))){
    data1 <- read.delim(file = paste(dir_data, "3.2.6_input_", strain[i], "_summary.txt", sep = ""),   header = T) # load values of fold change
and DEG classification
    fc_target1 <- data1[grep(paste(target, collapse = "|"), data1[,7]), ]
    a <- fc_target1
    if (nrow(fc_target1)>0){
      fc_target1 <- fc_target1[order(fc_target1[,4]), ]
      fc <- fc_target1[,4]
      deg <- fc_target1[,5] %>% as.character()
      out_fc[c(1:length(fc)), i] <- fc
      out_deg[c(1:length(fc)), i] <- deg
    }
  }
  out_fc <- out_fc %>% as.matrix()
```

```
out_deg <- out_deg %>% as.matrix()
z_fc <- melt(out_fc)
z_deg <- melt(out_deg)

#Step2: plot using ggplot2
optns <- theme(
  axis.text.x = element_text(angle = 270,hjust = 0, vjust = 0.5,margin = margin(c(20, 0, 0, 0)), size = 50, face = "bold", family =
"Helvetica"),
  axis.text.y = element_text(margin = margin(c(0, 20, 0, 0)) ,size = 50, face = "bold", family = "Helvetica"),
  axis.title = element_text(size = 50, face = "bold", family = "Helvetica"),
  axis.title.x = element_text(margin = margin(c(0, 0, 0, 0))),
  axis.title.y = element_text(margin = margin(c(0, 0, 0, 0))),
  axis.ticks.y = element_line(size = 3),
  axis.ticks.x = element_line(size = 3),
  axis.ticks.length = unit(.5, "cm"),
  axis.line    = element_line(size = 3),
  panel.background = element_blank(),
  legend.text = element_text(size = 40, face = "bold", family = "Helvetica", margin = margin(c(20, 0, 20, 0))),
  legend.title = element_text(size = 40, face = "bold", family = "Helvetica"),
  plot.title = element_text(hjust = 0.5, size = 20),
  plot.margin = margin(20, 0, 0, 0)
 )
 #making strain-color list
 commensal <- read.delim(file = "data/input/3.2.6/3.2.6_input_commensal_phylogeny.txt",   header = F, row.names = 1)
 strains <- c("Leaf404", "Leaf130", "Leaf155", "Leaf177", "Root935" ,"Leaf176", "Leaf187","Leaf1","Soil763", "Pto", "D36E","MM") %>%
as.data.frame()
 rownames(strains) <- strains[,1]
 strains[c(11:12), 1] <- "Pto"
 strains[,1] <- commensal[as.character(strains[,1]), 2]
 names(strains) <- "strains"
 list_strains <- list(strains = c(Actinobacteria ="firebrick3",
                                  Bacteroidetes = "blue",
                                  Firmicutes = "orange",
```

```
                            Alphaproteobacteria = "darkolivegreen4",
                            Betaproteobacteria = "darkolivegreen1",
                            Gammaproteobacteria = "darkgreen" )) %>%data.frame() # match different colors to different
commensal strains

 strain_exist <- z_fc[!is.na(z_fc[, 3]), 2] %>% as.character()
 strain_melt <-z_fc[, 2] %>% as.character()
 phyla_melt <- strains[strain_melt , 1]%>% as.factor()
 phyla_exist <- strains[strain_exist , 1]%>% as.factor()

 #determine colors for phyla, in the order appears in the plot (alphabetic)
 strain_color2 <- list_strains[as.character(sort(unique(phyla_exist))), 1] %>% as.character()

 #for coloring DEGs
 deg_color <- z_deg[,3]
 deg_color <- gsub("nonDEG", "black", deg_color)
 deg_color <- gsub("down", " deepskyblue3", deg_color)
 deg_color <- gsub("up", "deeppink", deg_color)%>% na.omit() %>% as.character()


 q <- ggplot(z_fc, aes(x = Var2, y = value,   color= phyla_melt)) +
   geom_boxplot(outlier.color = NA, size = 1.5, fill = NA) +
   scale_fill_grey() +
   theme_classic() +
   optns +
   scale_color_manual(values = unique(strain_color2)) +
   geom_dotplot(binaxis = 'y', stackdir = 'centerwhole', position = position_jitter(.1),
                binwidth = .1 , alpha = 1, dotsize = 2,   fill = deg_color, color = NA) +
   xlab("") + ylab("") + theme(legend.position = "none") +
   scale_x_discrete(limits = strain2) +
   scale_y_continuous(labels = scales::number_format(accuracy = 0.1, decimal.mark = ".")) #use the same n of decimal
 q
}
```

With the following R script, gene expression can be visualized iteratively.

```
strain <- c("Leaf1","Leaf130", "Leaf155","Leaf177", "Leaf187", "Leaf176", "Leaf404",
"Root935", "Soil763", "Pto", "D36E", "MM")
dir_database <- "data/input/3.2.6/"
out_dir <- "/data/output/3.2.6/"


#loading the KO database
ko_database <- read.delim(file = paste(dir_database, "3.2.6_input_KO_master_database.txt",
sep = ""),   header = F)


#search for KO pathway annotations
infile <- read.delim(file = paste(dir_database,
"3.2.6_input_ko_pathway_annotation_list_for_plots.txt", sep = ""),   header = T)
ko2 <- infile[, 1] %>% unique()
for (i in c(1:length(ko2))) {
  l <- ko_database[grep(ko2[i], ko_database[, 4]), 1] %>% unique() %>% as.character()
  if (length(l)>0){
    q <- ko.plot(l)
  }
  prefix <- strsplit(ko2[i], '\\\\')[[1]][1]
  ggsave(paste0(out_dir, "Expression_boxplot_categ3_", prefix,".png"),   height = 10, width =
13, q)
}
```

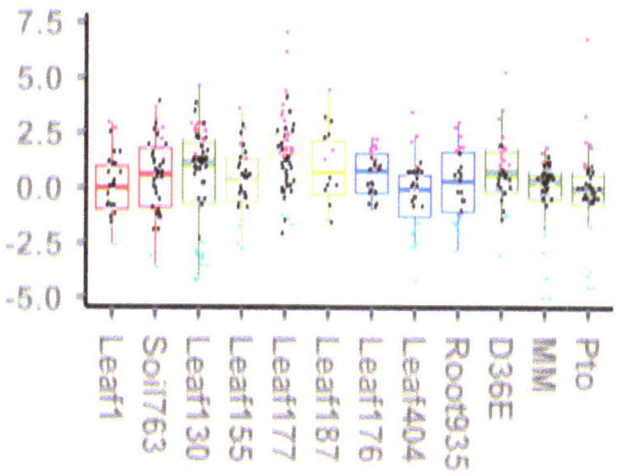

Fig. 16 A box plot for KO terms expression visualization

You can visualize the expression of selected KO terms by passing them as a vector to ko.plot():

```
target <- c("K15553|K15554|K15555|K02047|K02048")

q <- ko.plot(target)

ggsave(paste(out_dir, "Expression_boxplot_sulfur_transport.png"),  height = 10, width =
13, q)
```

These visualizations enable a high-level comparison of functional gene expression dynamics across bacterial taxa and treatment conditions, highlighting conserved or strain-specific transcriptional responses (*see* Fig. 16).

3.3 Integration of Plant and Bacterial Transcriptomes

3.3.1 Identification of Plant-Associated Genes

To identify bacterial genes associated with plant interactions, we focused on plant-associated (PA) genes, defined as genes annotated with KO terms that are classified as PA by Levy et al. [19]. The full list of plant-associated KO terms is available at: GitHub https://github.com/tnobori/co-transcriptomics/tree/master/data/plant-associated_genes. Only KO terms marked as "significant" in at least two independent statistical tests were retained. Genes with these KO terms were labeled as "PA"; others were labeled as "nonPA."

As an example, the following R script identifies plant-associated genes in the Leaf1 bacterial sample:

```
data <- read.delim("data/input/3.3.1/3.3.1_input_Leaf1_summary.txt", header = T, row.names =
1)
pa_threshold = 1
dat <- read.delim("data/input/3.3.1/3.3.1_input_ko_Actinobacteria.txt", header = T, row.names
= 1) # Statistical test results of KO terms
rownames(dat) <- gsub("KO:", "", rownames(dat))
dat <- dat[, -c(1:7)]

pa_test <- apply(dat, 1, function(x){sum(x=="Y")}) #number of statistical tests that were
significant
pa <- names(pa_test)[pa_test > pa_threshold] %>% as.character() #significant in at least two
statistical test

data[, ncol(data)+1] <- "nonPA"
b <- grep(paste(pa, collapse = "|"), data[, 6]) %>% na.omit() %>% as.numeric()
data[b,ncol(data)] <- "PA"
colnames(data)[ncol(data)] <- "Plant Associated"

data_for_save <- data[, c(5,6,7,1,2,3,4,8)]
write.table(data_for_save, file="data/output/3.3.1/
3.3.1_output_gene_expression_withPA_Leaf1.txt", row.names=T, col.names=NA, sep="\t",
quote=F)
```

	OG	KO.id	KO.description	L1_Col	L1_TSB	L1_FC	DEG	Plant Associated
cds100	OG0001554	K13788	pta; phosphate acetyltransferase [EC:2.3.1.8]	10.2060777	9.60259873	1.19953431	up	nonPA
cds1000	OG0000015	K02031	ABC.PE.A; peptide/nickel transport system ATP-binding protein	4.21018307	5.46611822	-0.4641249	nonDEG	nonPA
cds1001	OG0000104	K16568	exoZ; exopolysaccharide production protein ExoZ	9.88964167	11.0215995	-0.5323686	nonDEG	nonPA
cds1003	OG0000109			5.56071849	6.05873011	-1.7474905	nonDEG	nonPA
cds1004				9.29732274	7.40027192	2.51679017	up	nonPA
cds1005	OG0000374	K01414	prlC; oligopeptidase A [EC:3.4.24.70]	11.4679969	10.7914023	1.28207858	up	nonPA
cds1006	OG0000705	K03544	clpX; ATP-dependent Clp protease ATP-binding subunit ClpX	8.36145748	7.81783292	1.182902	nonDEG	nonPA
cds1007	OG0000451	K01358	clpP; ATP-dependent Clp protease, protease subunit [EC:3.4.21.92]	11.1978842	10.3374165	1.47874947	up	nonPA
cds1008	OG0000451	K01358	clpP; ATP-dependent Clp protease, protease subunit [EC:3.4.21.92]	4.08093222	6.30372495	-1.7711167	nonDEG	nonPA
cds1009				8.7854846	10.7322994	-1.3446739	down	nonPA
cds1012	OG0000373			12.3513075	9.7210281	3.25304865	up	nonPA
cds1013	OG0000477	K04047	dps; starvation-inducible DNA-binding protein	5.77709093	6.97240818	-0.5468268	nonDEG	nonPA
cds1014	OG0000176			7.61405718	7.98854976	0.26446838	nonDEG	nonPA
cds1015	OG0000461	K10563	mutM; formamidopyrimidine-DNA glycosylase [EC:3.2.2.23 4.2.99.18	6.96365924	8.19701037	-0.4234925	nonDEG	nonPA
cds1016	OG0001887	K01808	rpiB; ribose 5-phosphate isomerase B [EC:5.3.1.6]	8.79975869	8.30984758	1.08230791	nonDEG	nonPA

Fig. 17 A partial of the output, 3.3.1_output_gene_expression_withPA_Leaf1

This outputs a file (3.3.1_output_gene_expression_withPA_Leaf1 .txt) that integrates DEG classification with PA status for each gene (*see* Fig. 17).

To compare transcriptional responses between PA and nonPA genes, the proportions of DEG categories (upregulated, downregulated, and non-DEG) can be calculated for each group:

```
data2 <- data[, c(6, 4, 8)]

data2 <- data2[!data2[, 1]=="", ]

out_pa <- table(data2[data2[,3]=="PA" ,2])

out_nonpa <- table(data2[data2[,3]=="nonPA" ,2])

out_pa2 <- out_pa/sum(out_pa)

out_nonpa2 <- out_nonpa/sum(out_nonpa)

names(out_pa2) <- c("PA down", "PA nonDEG","PA up")

names(out_nonpa2) <- c("nonPA down", "nonPA nonDEG","nonPA up")

print(out_pa2)

print(out_nonpa2)
```

Example Output: (*See* Fig. 18)

- PA genes: 3.5% down-regulated, 17.4% up-regulated
- nonPA genes: 7.6% down-regulated, 13.3% up-regulated

This analysis highlights functional biases in the regulation of PA versus nonPA genes during host colonization.

To visually compare expression profiles between PA and nonPA bacterial genes, a violin plot was generated using ggplot2 (*see* Fig. 19):

```
>   print(out_pa2)
   PA down  PA nonDEG        PA up
0.03548387 0.79032258 0.17419355
>   print(out_nonpa2)
  nonPA down nonPA nonDEG     nonPA up
   0.0755814    0.7913437    0.1330749
```

Fig. 18 The proportions of DEG categories for each group

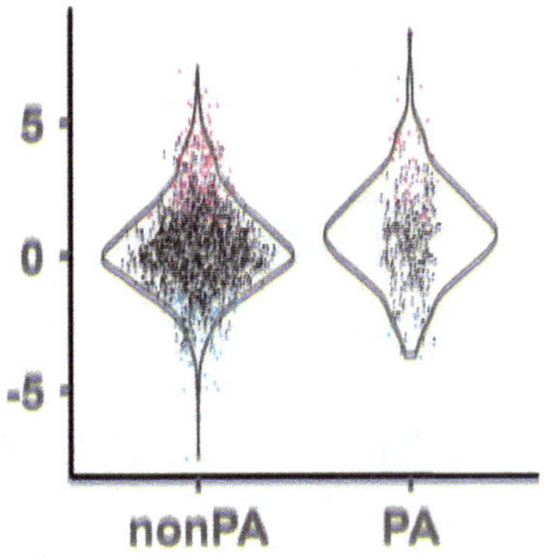

Fig. 19 3.3.1_output_Expression_boxplot_PA_DEG_Leaf1.png, A violin plot for PA and nonPA gene expression visualization

```
x <- data[, c(3, 4, 6, 8)]
x <- x[!x[, 3]== "", ]
x <- x[order(x[, 1]), ] %>% na.omit()
x <- x[order(x[, 4]), ] %>% na.omit()
colnames(x) <- c("FC", "DEG", "KO","PA")

q <- ggplot(x, aes(x = PA, y = FC)) + geom_violin( size = 1.5, fill = NA)
q <- q + scale_fill_grey() + theme_classic()
q <- q + optns


#setting DEG colors
c3 <- gsub("nonDEG", "black", x[, 2])
c3 <- gsub("down", "deepskyblue3", c3)
c3 <- gsub("up", "deeppink", c3)%>% na.omit() %>% as.character()


q <- q + geom_dotplot(binaxis = 'y', stackdir = 'centerwhole', position = position_jitter(.1),
binwidth = .1 , alpha = 1, dotsize = 1,   fill = c3, color = NA)+
xlab("") + ylab("")+ theme(legend.position = "none")

ggsave("data/output/3.3.1/3.3.1_output_Expression_boxplot_PA_DEG_Leaf1.png", height =
10, width = 10, q)
```

This plot illustrates the distribution of fold changes, with individual points colored by DEG classification.

To gain insight into the functional roles of PA versus nonPA genes under different regulatory states (e.g., upregulated, downregulated), KO term enrichment analysis can be performed for each subgroup using the ko.enrich.strain() function:

```
infile <- data2 %>% na.omit()
infile[, 2] <- paste(infile[, 2], infile[, 3], sep = "_")
pt <- unique(infile[, 2]) %>% as.character()
out <- data.frame("KO" = NA, "pval" = NA, "FoldEnrichment" = NA, "genes" = NA, "size" = NA,
"cluster" = NA)

#ko enrichment summary table
for (id in c(1:length(pt))){
  indata <- infile[infile[, 2]==pt[id] ,1]
  indata <- as.data.frame(indata)

  hypr <- ko.enrich.strain(indata, strain[i])
  if (nrow(hypr)>0){
    hypr[, 6] <- pt[id]
    colnames(hypr)[6] <- "cluster"
    out <- rbind(out, hypr)
  }
}
out <- out[-1, ]

write.table(out, file= "data/output/3.3.1/3.3.1_output_PA_DEG_Leaf1_KOenrich.txt",
row.names=F,   sep="\t", quote=F)
```

KO	pval	FoldEnrichment	genes	size	cluster
ABC transporters	1.48E-05	3.451084082	K17317---gtsC, glcG; glucose/mannose transport system p	13	up_nonPA
DNA repair and recombination proteins [BR:ko03400]	0	9.495586381	K03111---ssb, single-strand DNA-binding protein|K02337	15	up_nonPA
DNA replication proteins [BR:ko03032]	0	10.28688525	K03111---ssb, single-strand DNA-binding protein|K02337	5	up_nonPA
Enzymes with EC numbers	5.48E-09	5.809064609	K03207---wcaH; colanic acid biosynthesis protein WcaH [E(	12	up_nonPA
Exosome [BR:ko04147]	2.18E-08	5.094457455	K01897---ACSL, fadD; long-chain acyl-CoA synthetase [EC	13	up_nonPA
General function prediction only	0	13.71584699	K06876---K06876; deoxyribodipyrimidine photolyase-relat	5	up_nonPA
Glutathione metabolism	0	13.71584699	K00031---IDH1, IDH2, icd; isocitrate dehydrogenase [EC:1.1	5	up_nonPA
Homologous recombination	1.02E-05	6.857923497	K03111---ssb; single-strand DNA-binding protein|K02337	5	up_nonPA
Lipid biosynthesis proteins [BR:ko01004]	0	13.71584699	K01897---ACSL, fadD; long-chain acyl-CoA synthetase [EC	5	up_nonPA
Mismatch repair	0	41.14754098	K03111---ssb; single-strand DNA-binding protein|K02337	5	up_nonPA
Quorum sensing	0	18.51639344	K01897---ACSL, fadD; long-chain acyl-CoA synthetase [EC	9	up_nonPA
Transcription factors [BR:ko03000]	0	69.95081967	K09017---rutR; TetR/AcrR family transcriptional regulator|I	17	up_nonPA
Transporters [BR:ko02000]	0	16.06713505	K17317---gtsC, glcG; glucose/mannose transport system p	41	up_nonPA
Two-component system	0	41.14754098	K07687---rcsB; two-component system, NarL family, captu	10	up_nonPA
Two-component system [BR:ko02022]	0	9.875409836	K07687---rcsB; two-component system, NarL family, captu	6	up_nonPA
ABC transporters	0	2.806519297	K16785---ecfT; energy-coupling factor transport system pi	59	nonDEG_nonPA
Amino acid related enzymes [BR:ko01007]	0	2.38206527	K01885---EARS, gltX; glutamyl-tRNA synthetase [EC:6.1.1	21	nonDEG_nonPA
Amino sugar and nucleotide sugar metabolism	0	3.265211918	K00790---murA; UDP-N-acetylglucosamine 1-carboxyviny	31	nonDEG_nonPA

Fig. 20 A portion of the output, 3.3.1_output_PA_DEG_Leaf1_KOenrich.txt

This produces the file 3.3.1_output_PA_DEG_Leaf1_KOenrich.txt, which contains enriched KO terms along with *p*-values, fold enrichment, and gene lists (*see* Fig. 20).

This analysis allows the dissection of whether PA genes are functionally enriched for specific biological processes.

To statistically assess whether upregulated PA genes and downregulated nonPA genes were overrepresented, hypergeometric tests can be applied:

```
p <- phyper(q = out_pa[3], m = out_pa[3]+out_nonpa[3], n = sum(out_pa)+sum(out_nonpa)-
out_pa[ 3]-out_nonpa[3], k = sum(out_pa), lower.tail = F, log.p = F )
pp <- phyper(q = out_nonpa[1], m = out_pa[1]+out_nonpa[1], n =
sum(out_pa)+sum(out_nonpa)-out_pa[ 1]-out_nonpa[1], k = sum(out_nonpa), lower.tail = F,
log.p = F )
pp <- c(p, pp)
pp <- p.adjust(pp, method = "fdr") %>% as.data.frame()
print(pp)
```

Both tests yielded adjusted p-values <0.05, indicating significant enrichment of upregulated PA genes and downregulated nonPA genes. These findings support the hypothesis that plant-associated genes are preferentially activated during host interaction, while nonPA genes tend to be suppressed.

3.3.2 Integrated Correlation Analysis

This section outlines the preprocessing of both plant and bacterial transcriptomic data to prepare for cross-kingdom correlation analysis. Expression values are normalized and transformed into relative expression profiles to enable strain-level comparisons.

Plant Genes

1. Normalize raw read counts using edgeR.
2. Calculate average normalized expression per strain.
3. Compute gene-wise relative expression (mean-centered).

The input file, 3.3.2_input_plant_count.txt, includes all plant samples' raw gene counts processed followed by Subheadings 3.1.1, 3.1.2, and 3.1.3.

```
data <- read.csv("data/input/3.3.2/3.3.2_input_plant_count.txt",header = T,row.names = 1,
sep='\t')
data_ave5 <- data[apply(data, 1, mean)> 5, ]


# Step 1: normalized the raw reads count
dge <- DGEList(counts = data_ave5)
nf <- calcNormFactors(dge, logratioTrim = 0.3, sumTrim = 0.05)


lib.size <- colSums(data_ave5)*nf$samples$norm.factors


data_norm <- sweep(data_ave5, 2, lib.size, FUN="/")
data_norm <- data_norm*1000000
data_norm_log <- log(data_norm, base=2)


is.na(data_norm_log) <- sapply(data_norm_log, is.infinite)


# Step 2: Calculate the mean value of normalized expression for each gene and each plant
row.means <- rowMeans(data_norm_log)
split_str <- function(x){
  split.str <- strsplit(x, split='_')[[1]][1]
```

```
  return(split.str)
}
samples <- sapply(colnames(data_norm_log), split_str)
groups <- unique(factor(samples))
for (g in groups){
  data_norm_log$g <- apply(data_norm_log[which(samples==g)], 1, mean, na.rm=TRUE)
  colnames(data_norm_log)[length(colnames(data_norm_log))] = g
}
out <- data_norm_log[, unique(samples)]

# Step 3: calculate the relative expression level
row.means <- rowMeans(out)
rowMean <- function(x){
  x - mean(x, na.rm=TRUE)
}

out.row.mean <- t(apply(out, 1, rowMean))
write.csv(out.row.mean, file='data/output/3.3.2/3.3.2_output_plant_RNA-
seq_fitted_mean_DEG_atleast_one_strain.txt', sep='\t', row.names=T)
```

Bacterial OGs

1. Normalize expression data per strain.
2. Aggregate expression values by OG.
3. Transform into relative OG expression.

There are two inputs for this process for each strain. The first input is the original read counts data, 3.3.2_input_Leaf1_count.txt. The second input is the OG annotation file 3.3.2_input_annotation_comb_Leaf1.txt.

```
library(edgeR)

library(dplyr)

strain <- c("Leaf404", "Leaf130", "Leaf155", "Leaf177", "Root935" ,"Leaf176",

"Leaf187","Leaf1","Soil763")

abbre <- c('L404_col', 'L130_col', 'L155_col', 'L177_col', 'R935_col', 'L176_col', 'L187_col',

'L1_col', 'S763_col')

strain_abbre <- data.frame(strain=strain, abbre=abbre)

out.norm <- data.frame()


# Step 1: For each strain, generate the mean value of normalized expression of each gene

for (i in c(1:length(strain))){

    data <- read.table(paste(c('data/input/3.3.2/3.3.2_input_', strain[i], '_count.txt'),

collapse=''), header=T, sep='\t', row.names=1)

    data_ave5 <- data[apply(data, 1, mean)> 5, ]

    dge <- DGEList(counts = data_ave5)

    nf <- calcNormFactors(dge, logratioTrim = 0.3, sumTrim = 0.05)

    lib.size <- colSums(data_ave5)*nf$samples$norm.factors

    data_norm <- sweep(data_ave5, 2, lib.size, FUN="/")

    data_norm <- data_norm*1000000

    data_norm_log <- log(data_norm, base=2)

    is.na(data_norm_log) <- sapply(data_norm_log, is.infinite)


     row.means <- rowMeans(data_norm_log)
```

```
split_str <- function(x){
    split.str <- strsplit(x, split='_')[[1]]
    l <- length(split.str) - 1
    paste(split.str[0:l], collapse='_')
}
samples <- sapply(colnames(data_norm_log), split_str) # e.g. "Leaf1_vitro_1" into "
Leaf1_vitro "
col_id <- unique(samples)[! grepl('vitro', unique(samples))]
groups <- unique(factor(samples))
for (g in groups){
    data_norm_log$g <- apply(data_norm_log[which(samples==g)], 1, mean,
na.rm=TRUE)
    colnames(data_norm_log)[length(colnames(data_norm_log))] = g
}
out <- data_norm_log[, unique(samples)]
if (dim(out.norm)[1] == 0) {
    out.norm <- out
} else {
    out.norm <- merge(out.norm, out, by='row.names', all=TRUE)
    rownames(out.norm) <- out.norm[, 'Row.names']
    out.norm <- out.norm %>% select(-c(Row.names))
}
}
```

```
cds_id <- grepl('cds', rownames(out.norm))
out.norm <- out.norm[cds_id, unique(colnames(out.norm))[! grepl('vitro',
unique(colnames(out.norm)))]]
colnames(out.norm) <- strain_abbre[match(colnames(out.norm), strain_abbre$abbre), 'strain']
# Step 2: OG annotation and mean value calculation for each strain
out.og <- data.frame()
for (i in 1:length(strain)){
    tmp.og <- data.frame(); head <- c(); value <- c()
    OG_anno <- read.table(paste(c('data/input/3.3.2/3.3.2_input_annotation_comb_',
strain[i], '.txt'), collapse=''), header=T, sep='\t', row.names=1, quote='')
    for (og in unique(OG_anno$OG)){
        head <- c(head, paste(unique(OG_anno[OG_anno$OG == og,
c('OG','KO.id','KO.description')]), collapse='|'))
        tmp.cds <- OG_anno[OG_anno$OG == og, 'cds']
        value <- c(value, mean(out.norm[tmp.cds, strain[i]]))
    }
    tmp.og <- data.frame(HEAD=head, VALUE=value)
    colnames(tmp.og) <- c('OG_ID', strain[i])
    if (dim(out.og)[1] == 0){out.og <- tmp.og
    } else {out.og <- merge(out.og, tmp.og, by='OG_ID', all=TRUE)
    }
}
rownames(out.og) <- out.og[, 'OG_ID']
```

```
out.og <- out.og %>% select(-c('OG_ID'))

# Step 3: Calculate the relative expression level of each OG

row.means <- rowMeans(out.og)

rowMean <- function(x){

  x - mean(x, na.rm=TRUE)

}

out.row.mean <- t(apply(out.og, 1, rowMean))


write.table(out.row.mean, 'data/output/3.3.2/3.3.2_output_bacteria_RNA-seq_combined.txt',

sep='\t', row.names=T, quote=F)
```

This completes the data normalization and transformation for both plant and bacterial datasets, resulting in relative expression matrices suitable for integrated correlation analysis.

To facilitate cross-kingdom correlation analysis, plant gene expression values were converted to fold-change relative to mock-treated samples. This normalization step ensures comparability across strains by expressing gene activity as treatment-specific deviations from the mock baseline:

```
strains <- c("Leaf404", "Leaf130", "Leaf155", "Leaf177", "Root935" ,"Leaf176",
"Leaf187","Leaf1","Soil763")

plant <- read.delim("data/output/3.3.2/3.3.2_output_plant_RNA-
seq_fitted_mean_DEG_atleast_one_strain.txt", header = T, row.names = 1) #plant RNA-seq
data; genes that are DEG at least in one of the strains

a <- grep("mock", colnames(plant)) #mock
plant2 <- sweep(plant, 2, plant[, a]) #fold changes
plant2 <- plant2[, -a]
plant2 <- plant2[, colnames(bac)]
```

To ensure robust estimation of correlations between plant DEGs and bacterial OGs, a bootstrapped correlation strategy is employed. This approach minimizes the influence of any single strain and identifies consistently strong associations:

1. Leave-one-out correlations: Compute correlation matrices with each strain removed in turn.
2. Full correlation: Compute the correlation matrix using all strains.
3. Stability filtering: For each gene pair, retain the minimum absolute correlation across all runs if the sign is consistent; otherwise, assign 0.

```
strains <- c("Leaf404", "Leaf130", "Leaf155", "Leaf177", "Root935" ,"Leaf176",
"Leaf187","Leaf1","Soil763")
bac <- read.delim("data/output/3.3.2/3.3.2_output_bacteria_RNA-seq_combined.txt", header
= T, row.names = 1) # bacterial OG expression file
bac <- bac[, -c(10:13)]
bac2 <- na.omit(bac)
#Step1: function to calculate correlation without one strain
make_corr <- function(strain){
  #selected strains
  plant3 <- plant2[, strains[-grep(strain, strains)]]
  bac3 <- bac2[, strains[-grep(strain, strains)]]

  out1 <- matrix(data=NA, nrow = nrow(bac3), ncol = nrow(plant3),
                 dimnames = list(rownames(bac3), rownames(plant3)))

  for (i in c(1:nrow(plant3))){
    x <- apply(bac3, 1, function(x){cor(x,as.numeric(plant3[i, ]) )})
    out1[, i] <- x
  }
  out1
}

out1 <- make_corr(strains[1])
```

```
out2 <- make_corr(strains[2])
out3 <- make_corr(strains[3])
out4 <- make_corr(strains[4])
out5 <- make_corr(strains[5])
out6 <- make_corr(strains[6])
out7 <- make_corr(strains[7])
out8 <- make_corr(strains[8])
out9 <- make_corr(strains[9])


#Step2: calculate correlation using all strains
plant3 <- plant2
bac3 <- bac2
out <- matrix(data=NA, nrow = nrow(bac3), ncol = nrow(plant3),
                    dimnames = list(rownames(bac3), rownames(plant3)))
for (j in c(1:nrow(plant3))){
   x <- apply(bac3, 1, function(x){cor(x,as.numeric(plant3[j, ]) )})
   out[, j] <- x
}
```

```
#Step3: take the smallest coefficients as the final correlation coefficients
out_integrated <- out
out_integrated[] <- 0

for (i in c(1:nrow(out))){
  for (j in c(1:ncol(out))){
    if (out[i,j]>0 & out1[i,j]>0 & out2[i,j]>0 & out3[i,j]>0 & out4[i,j]>0 & out5[i,j]>0 & out6[i,j]>0 &
out7[i,j]>0 & out8[i,j]>0 & out9[i,j]>0){
      out_integrated[i, j] <- min(out[i,j], out1[i,j], out2[i,j], out3[i,j], out4[i,j], out5[i,j], out6[i,j],
out7[i,j], out8[i,j], out9[i,j])
    } else if (out[i,j]<0 & out1[i,j]<0 & out2[i,j]<0 & out3[i,j]<0 & out4[i,j]<0 & out5[i,j]<0 &
out6[i,j]<0 & out7[i,j]<0 & out8[i,j]<0 & out9[i,j]<0){
      out_integrated[i, j] <- max(out[i,j], out1[i,j], out2[i,j], out3[i,j], out4[i,j], out5[i,j], out6[i,j],
out7[i,j], out8[i,j], out9[i,j])
    } else{
      out_integrated[i ,j] <- 0
    }
  }
}

write.table(out_integrated, file=
"data/output/3.3.2/3.3.2_output_plantDEG_bac_correlation.txt", row.names=T, col.names=NA,
sep="\t", quote=F)
```

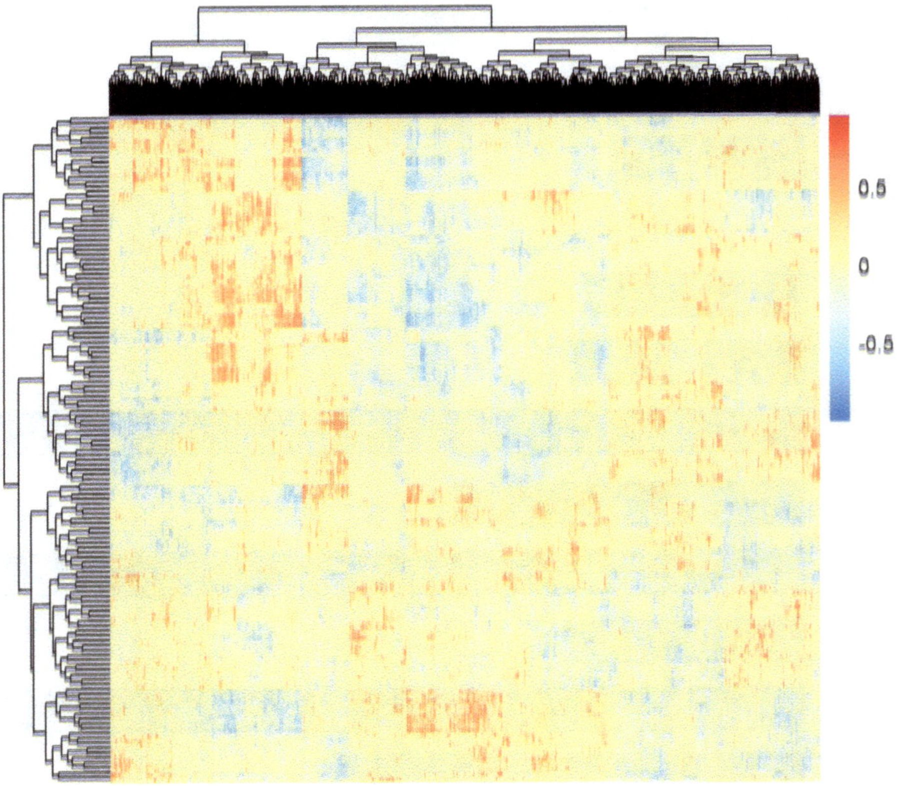

Fig. 21 A heatmap for generated correlation matrix visualization

The generated correlation matrix (3.3.2_output_plantDEG_bac_correlation.txt) can be visualized using pheatmap: (*see* Fig. 21).

```
pheatmap(out_integrated, show_rownames=F, show_colnames=F)
```

4 Notes

1. For users without access to a high-performance Linux environment, we recommend using the web-based tool GEO2R (https://www.ncbi.nlm.nih.gov/geo/geo2r/) to process raw sequencing data hosted in the GEO database. GEO2R also allows for basic differential expression analysis directly through a web browser and serves as a practical entry point for transcriptome analyses.
2. Users may use gene count tables generated via GEO2R. Alternatively, preprocessed data tables can be downloaded from the project's GitHub repository. However, for consistency and

reproducibility, we recommend executing R scripts on a remote Linux server using the provided singularity container, which includes all required R packages and dependencies.

3. Detailed codes and processed data for this workflow can be accessed via https://github.com/YTang906/co-transcriptome.

References

1. Wang Z, Gerstein M, Snyder M (2009) RNA-Seq: a revolutionary tool for transcriptomics. Nat Rev Genet 10:57–63
2. Nakagami S, Wang Z, Han X, Tsuda K (2024) Regulation of bacterial growth and behavior by host plant. Annu Rev Phytopathol 62:69–96
3. Nobori T, Velásquez AC, Wu J, Kvitko BH, Kremer JM, Wang Y, He SY, Tsuda K (2018) Transcriptome landscape of a bacterial pathogen under plant immunity. Proc Natl Acad Sci USA 115:E3055–E3064
4. Entila F, Han X, Mine A, Schulze-Lefert P, Tsuda K (2024) Commensal lifestyle regulated by a negative feedback loop between Arabidopsis ROS and the bacterial T2SS. Nat Commun 15:456
5. Nobori T, Wang Y, Wu J, Stolze SC, Tsuda Y, Finkemeier I, Nakagami H, Tsuda K (2020) Multidimensional gene regulatory landscape of a bacterial pathogen in plants. Nat Plants 6: 883–896
6. Nobori T, Cao Y, Entila F, Dahms E, Tsuda Y, Garrido-Oter R, Tsuda K (2022) Dissecting the cotranscriptome landscape of plants and their microbiota. EMBO Rep 23:e55380
7. Nobori T, Tsuda K (2018) In planta transcriptome analysis of pseudomonas syringae. Bio Protoco 8:e2987–e2987
8. Bai Y, Müller DB, Srinivas G, Garrido-Oter R, Potthoff E, Rott M, Dombrowski N, Münch PC, Spaepen S, Remus-Emsermann M, Hüttel B, McHardy AC, Julia A, Vorholt JA, Schulze-Lefert P (2015) Functional overlap of the Arabidopsis leaf and root microbiota. Nature 528:364–369
9. Kim D, Paggi JM, Park C, Bennett C, Salzberg SL (2019) Graph-based genome alignment and genotyping with HISAT2 and HISAT-genotype. Nat Biotechnol 37:907–915
10. Anders S, Pyl PT, Huber W (2015) HTSeq—a Python framework to work with high-throughput sequencing data. Bioinformatics 31:166–169
11. Danecek P, Bonfield JK, Liddle J, Marshall J, Ohan V, Pollard MO, Whitwham A, Keane T, McCarthy SA, Davies RM, Li H (2021) Twelve years of SAMtools and BCFtools. GigaScience 10:giab008
12. Yu G (2024) Thirteen years of clusterProfiler. Innovation 5:100722
13. Chen Y, Chen L, Lun ATL, Baldoni P, Smyth GK (2025) edgeR v4: powerful differential analysis of sequencing data with expanded functionality and improved support for small counts and larger datasets. Nucleic Acids Res 53:gkaf018
14. Wickham H (2016) ggplot2: elegant graphics for data analysis. Springer-Verlag, New York
15. Slowikowski K (2024) Ggrepel: automatically position non-overlapping text labels with 'ggplot2'. https://ggrepel.slowkow.com/
16. Kolde R (2019) Pheatmap: pretty Heatmaps. R package version 1.0.12. https://CRAN.R-project.org/package=pheatmap
17. Storey JD, Bass AJ, Dabney A, Robinson D (2025) qvalue: Q-value estimation for false discovery rate control. R package version 2.40.0. https://bioconductor.org/packages/qvalue
18. Neuwirth E (2022) RColorBrewer: ColorBrewer palettes. R package version 1.1-3. https://CRAN.R-project.org/package=RColorBrewer
19. Levy A, Salas Gonzalez I, Mittelviefhaus M, Clingenpeel S, Herrera Paredes S, Miao J, Wang K, Devescovi G, Stillman K, Monteiro F, Alvarez BR, Lundberg DS, Lu TY, Lebeis S, Jin Z, McDonald M, Klein AP, Meghan E Feltcher ME, Rio TG, Grant SR, Doty SL, Ley RE, Zhao B, Venturi V, Pelletier DA, Vorholt JA, Tringe SG, Woyke T, Dangl JL (2018) Genomic features of bacterial adaptation to plants. Nat Genet 50:138–150

Chapter 14

Nuclei Isolation from Mature Leaves of *Arabidopsis thaliana* Infected by Bacterial Pathogens for Single-Nucleus Multi-Omics

Tatsuya Nobori

Abstract

Single-nucleus omics technologies are increasingly applied in plant biology to investigate transcriptional and chromatin regulation at cellular resolution. A key requirement for these approaches is the isolation of high-quality nuclei, which can be particularly challenging in plant tissues due to cell wall complexity and residual debris. Here, I present a streamlined protocol for isolating nuclei from fresh *Arabidopsis thaliana* leaves infected with the bacterial pathogen *Pseudomonas syringae*. This method eliminates the need for density gradient separation and fluorescence-activated nuclei sorting (FANS), reducing hands-on time while maintaining compatibility with single-nucleus RNA and ATAC sequencing using the 10x Genomics platform.

Key words Nuclei isolation, Single-cell transcriptomics, Single-cell epigenomics, Plant immunity, *Arabidopsis*

1 Introduction

Single-cell and single-nucleus omics technologies have transformed biological research by uncovering the molecular diversity of individual cells and their spatial organization within tissues [1]. These approaches are increasingly used in plant science to construct cell atlases, identify rare or transient cell states, and dissect tissue-specific responses to developmental and environmental cues [2, 3].

Nuclei isolation is a critical step for single-nucleus omics experiments. In plants, various protocols have been developed for different tissue types, using either fresh or frozen samples [4–6]. These protocols are often adapted from bulk omics workflows and typically involve filtration, density gradient separation, and fluorescence-activated nuclei sorting (FANS). Although FANS improves nuclei purity, it is time-consuming—particularly when extracts contain abundant non-nuclear debris such as cell wall

Tsutomu Kawasaki and Hirofumi Yoshioka (eds.), *Plant Immunity: Methods and Protocols*, Methods in Molecular Biology, vol. 3012,
https://doi.org/10.1007/978-1-0716-5138-4_14,

fragments. This poses a challenge for processing multiple samples in parallel, which is essential for cost-effective use of commercial single-cell platforms.

Here, I describe a simplified protocol optimized for mature *Arabidopsis thaliana* leaves. This method was successfully applied for single-nucleus multi-omics (snRNA-seq and snATAC-seq) analysis of pathogen-infected tissue using the 10x Genomics platform [7]. This protocol omits both density gradient separation and FANS, significantly reducing experimental time. It includes pathogen infection, tissue harvesting, mechanical disruption in the nuclei isolation buffer, debris filtration, and nuclei counting. For downstream library preparation and data analysis, we refer readers to the protocols of specific kits and a recent publication that provides detailed code and workflow examples [7].

2 Materials

1. King's B (KB) Medium: In a beaker or flask, add approximately 800 mL of distilled water. While stirring, add 20 g of Proteose Peptone No. 3, 1.5 g of K_2HPO_4, and 1.5 g of $MgSO_4{\cdot}7H_2O$. Add 10 mL of glycerol. Continue stirring until all components are fully dissolved. After all components dissolve, adjust the volume to 1 L with distilled water, then dispense the medium into suitable bottles or flasks, filling each no more than two-thirds full. Autoclave at 121 °C for 15 min.
2. Nuclei isolation buffer (NIB): Dissolve spermine and spermidine powders separately in nuclease-free water to prepare 0.2 M spermine and 0.5 M spermidine stock solutions (*see* **Note 1**). Store at –20 °C. For 50 mL NIB, mix the following: 750 μL of 1 M Tris–HCl (pH 7.5), 200 μL of 0.5 M EDTA, 2500 μL (2.5 mL) of 2 M KCl, 200 μL of 5 M NaCl, 50 μL of spermine and spermidine stocks (1:1000 each; final 0.2 mM spermine, 0.5 mM spermidine), and 500 μL of protease inhibitor cocktail (1:100). Add nuclease-free water to a final volume of 50 mL, mix by inversion, and keep on ice (*see* **Note 2**).
3. NIB-A: Mix 1 mL of NIB with RNase inhibitor at 1:100 (e.g., 10 μL RNaseIN per 1 mL NIB).
4. NIB-B: Mix 20 mL NIB, 1:500 RNaseIN (40 μL), and 10% BSA at 1:10 (10% solution) (2 mL added to reach ≈1% final).
5. NIB-C: Mix 20 mL NIB with 200 μL each of 10% Triton X-100 and 10% NP-40, RNase inhibitor at 1:1000 (e.g., 20 μL RNaseIN), and 10% BSA at 1:10 (2 mL).
6. NIB-D: Mix 10 mL NIB with RNase inhibitor at 1:1000 (e.g., 10 μL RNaseIN) and 10% BSA at 1:10 (2 mL).

7. Nuclei suspension buffer: Following the protocol of 10x Genomics Multiome, mix the following: 1000 μL of 1× Nuclei Buffer, RNase inhibitor at 1:40 (e.g., 25 μL RNaseIN), 1 mM DTT (1 μL of 1 M DTT per 1 mL total).
8. Swinging-bucket rotor centrifuge.

3 Methods

3.1 Infection of Bacterial Pathogens into Arabidopsis *Leaves*

1. Grow *Arabidopsis thaliana* (ecotype Col-0) on soil at 22 °C with 60% humidity and 12 h light cycles for approximately 28 days (*see* **Note 3**).
2. Three days before infection, streak *Pseudomonas syringae* onto King's B (KB) containing 1% Bacto Dehydrated Agar supplemented with appropriate antibiotics and incubate at 28 °C.
3. After two days, verify that discrete colonies have formed. Pick approximately 10 colonies into a 1.5 mL microcentrifuge tube containing 1 mL of liquid KB. Vortex thoroughly to mix.
4. Prepare 2–3 fivefold serial dilutions by adding 200 μL of the culture to 800 μL of KB medium.
5. Inoculate 50 μL of each dilution into separate test tubes containing 2 mL of KB with appropriate antibiotics. Incubate overnight (~16 h) at 28 °C with shaking.
6. Use a culture that has not reached saturation (*see* **Note 4**).
7. Transfer up to 2 mL of the overnight culture to a 2 mL microcentrifuge tube and centrifuge at 5200× *g* for 2 min at room temperature to pellet the bacteria.
8. Discard the supernatant and add 1 mL of sterile water. Vortex thoroughly to resuspend the pellet.
9. Repeat the centrifugation and wash steps once more.
10. After the final wash, resuspend the bacterial pellet in 1 mL of sterile water. Measure the OD_{600} and adjust the suspension to the desired concentration with sterile water.
11. Using a needle-less syringe, infiltrate the bacterial suspension into the abaxial (lower) side of fully expanded leaves (*see* **Note 5**).
12. Gently blot the leaf surface with a paper towel to remove excess liquid.

3.2 Nuclei Isolation

1. Pre-cool a swinging-bucket rotor centrifuge to 4 °C.
2. Keep all buffers (NIB-A, NIB-B, NIB-C, NIB-D, and Nuclei Suspension Buffer) chilled on ice or at 4 °C.
3. Using forceps, collect approximately 20 *P. syringae*-infected leaves into a Petri dish on ice (*see* **Notes 6** and 7).

4. Add 1 mL of NIB-A to the leaves and chop with a razor blade for 3–5 min until the tissue is homogenized.
5. Transfer the homogenate to a 50 mL conical tube on ice and add 20 mL of NIB-B.
6. (Optional) If processing multiple samples, repeat the above steps for each sample (*see* **Note 8**).
7. Filter the crude extract sequentially through a 70 μm and then a 30 μm mesh to remove large debris.
8. Add Triton X-100 and NP-40 (from 10% stocks) to a final concentration of 0.1% each (i.e., 200 μL of each detergent per 20 mL extract).
9. Incubate at 4 °C for 2 min with gentle rocking or rotation.
10. Centrifuge the sample at 50× *g* for 3 min at 4 °C in a swinging-bucket rotor to pellet non-nuclear debris (*see* **Note 9**).
11. Carefully transfer the supernatant to a new 50 mL tube on ice.
12. Centrifuge at 500× *g* for 5 min at 4 °C to pellet nuclei (*see* **Note 10**).
13. Discard the supernatant and gently resuspend the pellet in 1 mL of NIB-C by pipetting up and down ~10 times using a wide-bore tip (*see* **Note 11**).
14. Add the remaining 19 mL of NIB-C and incubate at 4 °C for 10 min with gentle agitation.
15. Centrifuge again at 500× *g* for 5 min at 4 °C to pellet nuclei.
16. If the pellet appears green, repeat the NIB-C wash once more (*see* **Note 12**).
17. Gently resuspend the pellet in 20 mL of NIB-D.
18. Centrifuge at 500× *g* for 5 minutes at 4 °C to pellet nuclei.
19. Discard the supernatant.
20. Add 950 μL of Nuclei Suspension Buffer to the pellet, resuspend gently, and transfer to a 1.5 mL tube.
21. Centrifuge at 50× *g* for 3 min at 4 °C to remove residual debris. Transfer the supernatant to a new 1.5 mL tube (*see* **Note 13**).
22. Centrifuge again at 500× *g* for 5 min at 4 °C to pellet nuclei.
23. Carefully remove the supernatant, leaving ~10 μL of buffer (*see* **Note 14**).
24. Pipette slowly up and down using a standard-bore tip until the pellet is completely resuspended (*see* **Note 15**).
25. Take 1 μL of the nuclei suspension, dilute 1:10 in NIB (without detergents), and add propidium iodide to a final concentration of 1 μg/mL. Count nuclei using a hemacytometer or other

appropriate method. Assess nuclei quality based on debris contamination, aggregation (*see* **Note 16**), and structural integrity (*see* **Note 17**).

26. Adjust the nuclei concentration as needed for downstream applications.

4 Notes

1. Spermine and spermidine are hygroscopic, especially in powder form. If using powders, weigh them quickly and use fresh stock solutions to ensure accurate concentrations.
2. Prepare 50 mL of NIB per sample and set aside approximately 1 mL on ice for downstream steps (e.g., Step 25 of nuclei isolation).
3. The optimal plant age may vary depending on growth conditions.
4. Avoid using saturated (stationary-phase) bacterial cultures, as they often contain many dead cells and can result in inconsistent inoculum concentrations.
5. Perform infiltration gently, and avoid creating a tight seal between the leaf and the syringe, as this can damage the leaf. If you see circular marks on the leaves after infiltration, this indicates the leaf is damaged. To minimize damaging the tissue, which can affect how the plant responds, limit how many times the syringe contacts the leaf. For example, infiltrate the entire leaf from two positions at the bottom corners, rather than from multiple points.
6. Harvest leaves as quickly as possible to minimize changes to the transcriptome and epigenome.
7. Using more plant tissue does not always yield more nuclei. If more tissue is needed, consider splitting it into separate samples, process them independently, and combine them at the end.
8. When processing multiple samples, keep all tubes on ice. Depending on experience and setup, homogenizing one sample may take around 10 min.
9. A visible pellet may not be present. Proceed to the next step regardless.
10. To minimize mechanical damage to nuclei, limit the number of centrifugation and resuspension steps throughout the protocol.
11. Use gentle resuspension at every wash step unless stated otherwise. Avoid vigorous pipetting, which can shear nuclei.

12. If the pellet remains green after the second NIB-C wash, an additional wash may help, but repeated washes in NIB-C can increase the risk of nuclei damage. A light green color typically indicates chloroplast contamination, which may increase chloroplast DNA contamination in downstream ATAC-seq.
13. You may repeat this step until no visible pellet remains.
14. The final volume of the nuclei suspension depends on your assay. For 10x Genomics multiome, up to 5 μL is allowed.
15. The nuclei suspension may appear white due to starch from leaves, which looks like small particles under the microscope. These typically do not interfere with downstream applications, including 10x Genomics assays and likely other platforms.
16. Large debris or nuclei aggregates can be removed using Flowmi Cell Strainers (40 μm), but filtering may significantly reduce yield. To minimize nuclei loss during filtering, consider diluting the sample with 10–20 mL of NIB-B before filtration. Alternatively, increasing BSA concentration to up to 2% can help reduce aggregation.
17. To assess nuclei quality, follow the 10x Genomics guidelines (https://kb.10xgenomics.com/hc/en-us/articles/360020348651-How-can-I-assess-the-quality-of-my-nuclei-for-Single-Cell-ATAC-or-Single-Cell-Multiome-ATAC-GEX-Sequencing). Damaged nuclei often aggregate. To reduce damage, consider adjusting Triton-X and NP-40 concentrations and shortening incubation time. Using less tissue may also help.

Acknowledgments

This work was supported by a grant to T.N. from The Gatsby Charitable Foundation.

References

1. Nobori T (2025) Exploring the untapped potential of single-cell and spatial omics in plant biology. New Phytol. PMID: 40398874
2. Nolan TM, Shahan R (2023) Resolving plant development in space and time with single-cell genomics. Curr Opin Plant Biol 102444
3. Tenorio Berrío R, Dubois M (2024) Single-cell transcriptomics reveals heterogeneity in plant responses to the environment: a focus on biotic and abiotic interactions. J Exp Bot 75:5188–5203
4. Lee TA, Nobori T, Illouz-Eliaz N, Xu J, Jow B, Nery JR, Ecker JR (2023) A single-nucleus atlas of seed-to-seed development in Arabidopsis. bioRxiv:2023.03.23.533992
5. Marand AP, Chen Z, Gallavotti A, Schmitz RJ (2021) A cis-regulatory atlas in maize at single-cell resolution. Cell 184:3041–3055.e21
6. Li C, Wood JC, Deans NC, Jarrell AF, Martin D, Mailloux K, Wang YW, Robin Buell C (2022) Nuclei isolation protocol from diverse angiosperm species. bioRxiv:2022.11.03.515090
7. Nobori T, Monell A, Lee TA, Sakata Y, Shirahama S, Zhou J, Nery JR, Mine A, Ecker JR (2025) A rare PRIMER cell state in plant immunity. Nature 638:197–205

Chapter 15

Multimodal Analysis of Phytoalexin Synthesis in *Arabidopsis* by Mass Spectrometry Imaging and Fluorescent Microscopy

Shigeyuki Betsuyaku, Kaoru Nakagawa, and Koji Okuda

Abstract

Phytoalexins are plant secondary antimicrobial compounds that are rapidly and locally accumulated de novo upon pathogen attacks. They are strongly correlated with disease resistance; therefore, the timing and the location of their synthesis and accumulation have been explored transcriptionally and metabolically using various means separately. In this chapter, by focusing on the *Arabidopsis* camalexin (CA), we describe protocols for multimodal in situ detection of CA and elemental distribution, as well as the transcriptionally active region of its synthesis gene *PHYTOALEXIN DEFICIENT 3* (*PAD3*) within the same leaf sample challenged with a pathogen.

Key words Camalexin, Fluorescence imaging, MALDI-MSI, LA-ICP-MSI, Promoter reporter

1 Introduction

Phytoalexins are low-molecular-weight antimicrobial compounds with great structural diversity that rapidly accumulate de novo upon pathogen attacks [1]. Owing to the nature of antimicrobial compounds and their strong inducibility associated with disease resistance, they are considered key chemical defense determinants in plant-microbe interactions such as gene-for-gene relationships [2]. When gene-for-gene resistance, nowadays called effector-triggered immunity, is triggered, rapid localized cell death occurs at the site of infection, which is surrounded by a few layers of cells in which various defense-related genes involved in phytoalexin synthesis pathways are activated [3]. Such local induction of phytoalexins at the site of pathogen infection has been reported either indirectly or directly. The commonly used method is expression analysis of genes required for phytoalexin synthesis. Accumulation of particular mRNAs involved in phytoalexin synthesis around the site of infection of *Phytophthora megasperma* f. sp. *glycinea* was

Tsutomu Kawasaki and Hirofumi Yoshioka (eds.), *Plant Immunity: Methods and Protocols*, Methods in Molecular Biology, vol. 3012, https://doi.org/10.1007/978-1-0716-5138-4_15,

demonstrated in parsley leaves by in situ RNA hybridization [4] The use of promoter-reporter techniques helped to simplify the experimental procedures employed to visualize where particular phytoalexin synthesis genes are activated, as exemplified by a study of the *PHYTOALEXIN DEFICIENT 3* (*PAD3*) gene, which is required for the synthesis of camalexin (CA), the phytoalexin in *Arabidopsis thaliana* [5]. Although these analyses help to elucidate the regulatory mechanisms governing transcriptional regulation leading to phytoalexin synthesis, they cannot clarify if a particular phytoalexin accumulates in specific plant cells. The alternative method is direct analysis of phytoalexin accumulation in plant tissues. Taking advantage of sorghum deoxyanthocyanidins as pigmented compounds, conventional wide-field microscopy revealed that these phytoalexins accumulate in cells that a fungal pathogen *Colletotrichum graminicola* is about to penetrate, as well as in cells surrounding the penetrated cells [6]. Beyond visible phytoalexin molecules, mass spectrometry imaging (MSI) is emerging as a molecular imaging technique that directly visualizes the spatial distributions of various molecules in living organisms including plants without the need for labels or fluorescent probes [7]. Matrix-assisted laser desorption/ionization (MALDI)-MSI can be used to spatially map phytoalexins in plant tissues. Accordingly, MSI revealed the spatial distribution of CA on the surface of *Arabidopsis* leaves in response to treatments with various bacteria including the phytopathogen *Pseudomonas syringae* pv. *tomato* DC3000 (*Pst*) [8]. Although various techniques have been applied to elucidate the role of phytoalexins, the regulatory mechanisms underlying their abovementioned local inducibility have not been revealed. In addition, it remains to be determined where phytoalexins are synthesized and whether they are transported to other cells.

To address these issues, we developed a multimodal imaging pipeline to capture the activation domain of a phytoalexin synthesis gene and accumulation of the corresponding phytoalexin in the same leaf samples (Fig. 1). Focusing on CA, the phytoalexin in the model plant *Arabidopsis*, we first established transgenic *Arabidopsis* plants expressing eYFP fused to a nuclear localization signal (eYFP-NLS) under the control of the promoter of *PAD3* (pPAD3), which is a key gene in CA synthesis [5, 9, 10]. A leaf of the established pPAD3-eYFP-NLS plants was inoculated with an avirulent strain of *Pst* and then subjected to fluorescence imaging to visualize the pPAD3 active domain in the infected tissue (Fig. 2a). Subsequently, the distribution of CA was analyzed using MALDI-MSI (Fig. 2b). When combined with genetic means, comparative analysis of the pPAD3 active domain and spatial distribution of CA (Fig. 2c) improves our understanding of the spatial and temporal regulation of CA synthesis. Furthermore, laser ablation inductively coupled plasma (LA-ICP)-MSI was applied to the same leaf sample for

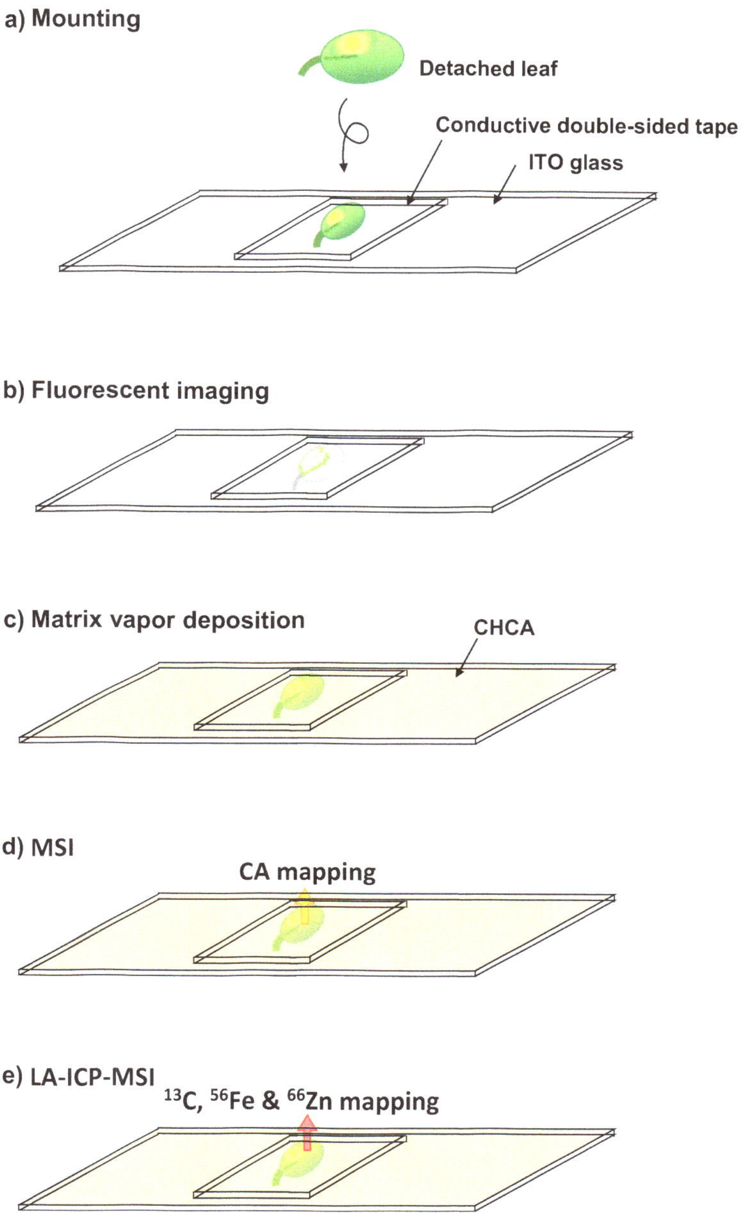

Fig. 1 Schematic representation of analysis of CA synthesis in *Arabidopsis*. (**a**) The pathogen-inoculated leaf of a pPAD3-eYFP-NLS plants is detached and mounted on the ITO glass using a piece of conductive double-sided tape. (**b**) Images of YFP and chlorophyll autofluorescence are acquired. (**c**) CHCA is vapor-deposited onto the surface of the specimen. (**d**) The distribution of CA is analyzed using MALDI-MSI. (**e**) The distributions of ^{13}C, ^{56}Fe, and ^{66}Zn are analyzed using LA-ICP-MSI

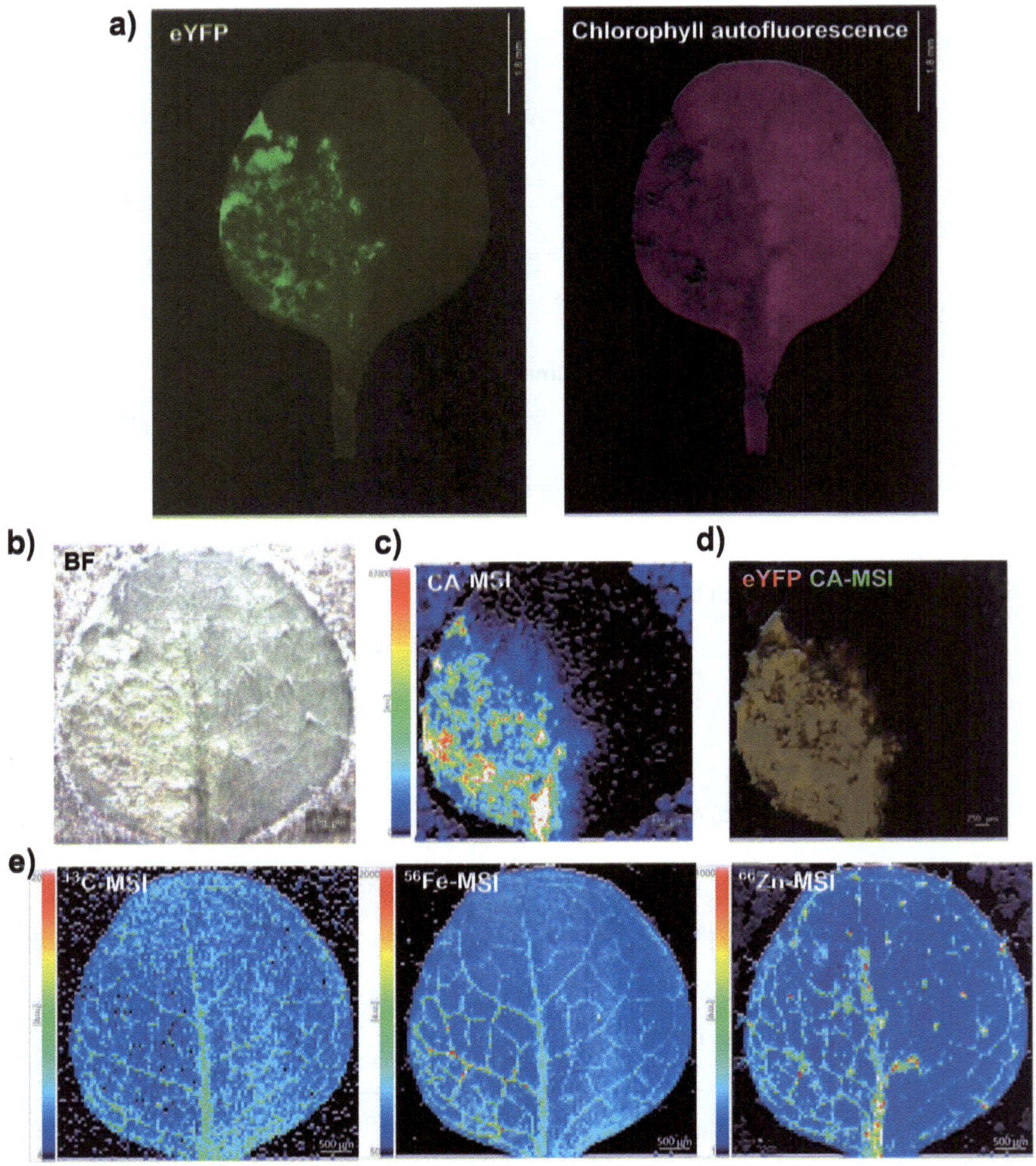

Fig. 2 Multimodal imaging analysis of CA synthesis in *Arabidopsis*. (**a**) Fluorescence stereomicroscopic images (left: eYFP, right: chlorophyll autofluorescence) of a pPAD3-YFP-NLS leaf partially infiltrated with *Pst_a2* ($OD_{600} = 0.02$) at 2 days post-infiltration. (**b**) Bright field image acquired using iMScope QT, (**c**) MALDI-MSI of the spatial distribution of CA in the same leaf as shown in **a**. (**d**) Overlaid image of the pPAD3 activity (red) and the distribution of CA (green). (**e**) LA-ICP-MSI of the spatial distributions of ^{13}C, ^{56}Fe, and ^{66}Zn in the same leaf

elemental spatial mapping of carbon (C), iron (Fe), and zinc (Zn). This revealed that Fe and Zn accumulated in the vasculature in the infiltrated area and that Zn specifically accumulated in trichomes (Fig. 2e). This combinatorial study might point to a functional relationship of these elements with CA and disease resistance. In this chapter, we describe the method for this multimodal imaging pipeline for CA synthesis in *Arabidopsis*. This method would also be useful to study the spatial and functional relationship between a particular metabolite and its synthesis gene, as well as various elements using living plant materials.

2 Materials

Prepare all solutions using ultrapure water and molecular biology grade reagents, unless otherwise stated. Autoclave the plant and bacterial media, and perform all procedures aseptically before sampling.

2.1 Preparation of Plants and Pathogens

2.1.1 Media

1. Plastic Petri plates (664,161, Greiner, Kremsmünster, Austria) for growing plants (*see* **Note 1**).
2. 1/2 Murashige and Skoog basal media containing 1% sucrose and 0.3% phytagel.
3. Mannitol-glutamate medium (10 g/L mannitol, 2.0 g/L L-glutamic acid, 0.5 g/L KH_2PO_4, 0.2 g/L NaCl, 0.2 g/L $MgSO_4/7H_2O$, and 15 g/L agar), supplemented with 100 mg/L rifampicin and 50 mg/L kanamycin.
4. PPM solution: 1% (v/v) Plant Preservative Mixture (Plant Cell Technology, Washington, DC, USA).
5. Surgical tape.

2.1.2 Preparation of Plants and Pathogens

1. Transgenic *Arabidopsis* plants carrying pPAD3-eYFP-NLS (*see* **Note 2**).
2. *Pst* carrying *AvrRpt2* (Pst_a2).
3. 10 mM $MgCl_2$.
4. Spectrophotometer.

2.2 Pathogen Inoculation

1. Latex gloves.
2. 1 mL needleless syringe (Terumo, Tokyo, Japan).

2.3 Fluorescence Microscopy and Mass Spectrometers

2.3.1 Preparation of Leaf Specimen for Fluorescence Microscopy and MSI

1. Conductive double-sided tape (AL-25DC; 3 M, Tokyo, Japan).
2. Indium tin oxide (ITO)-coated glass slide (SI0100N; Matsunami, Osaka, Japan).
3. iMLayer matrix vapor deposition system (Shimadzu, Kyoto, Japan).
4. α-Cyano-4-hydroxycinnamic acid (CHCA).

2.3.2 Fluorescence Microscopy

1. M205FA automated fluorescence stereomicroscope equipped with a DFC365FX camera and TXR and YFP filters (Leica Microsystems).

2.3.3 Mass Spectrometers

1. iMScope™ QT and LCMS-9050 (Shimadzu, Kyoto, Japan).
2. ICPMS-2030 (Shimadzu, Kyoto, Japan) as an ICP-MS, coupled with imageBIO266 (Elemental Scientific Lasers, Omaha, USA) as the LA system.

2.3.4 Data Analysis Software

1. IMAGEREVEAL™ MS (Shimadzu, Kyoto, Japan).

3 Methods

3.1 Preparation of Plants and Pathogens

1. Surface-sterilize the seeds of pPAD3-eYFP-NLS *Arabidopsis* plants by soaking them overnight in PPM solution at room temperature.
2. Sow the sterilized seeds onto 1/2 Murashige and Skoog basal media containing 1% sucrose and 0.3% phytagel.
3. Grow the seeds at 23 °C with 24 h of light for the indicated number of days.
4. Two days prior to pathogen inoculation, streak the bacterial pathogen *Pst_a2* from a glycerol stock onto mannitol-glutamate medium and incubate for 2 days at 28 °C.
5. Harvest the bacterial cells grown on the medium using sterilized plastic tips into a plastic tube containing 10 mM $MgCl_2$. Resuspend the cells in 10 mM $MgCl_2$ and adjust the optical density at 600 nm (OD_{600}) to 10^7 colony-forming units (cfu)/mL (normally corresponding to $OD_{600} = 0.02$).

3.2 Pathogen Inoculation

1. Select an appropriate leaf to be inoculated.
2. Wear disposable latex gloves prior to infiltration of the pathogen.
3. Open the lid of the Petri dish containing the growing plants in a laminar flow cabinet.
4. Carefully infiltrate the abaxial side of the selected leaf with the bacterial suspension at 10^7 cfu/mL (*see* **Note 3**).
5. Using a piece of soft paper towel, absorb the excess bacterial suspension on the infiltrated leaf.
6. Close and re-seal the lid with surgical tape.
7. Incubate the infiltrated plant at 23 °C with 24 h of light for a further 2 days.

3.3 Fluorescence Microscopy

To perform multimodal analysis using both fluorescence microscopy and MSI, the order of analysis is also important. Fluorescence imaging must be performed first, followed by MALDI-MSI, and finally LA-ICP-MSI. By attaching the leaf horizontally to the glass, it is easier to compare multiple images obtained using different imaging technologies.

1. Place a piece of conductive double-sided tape onto an ITO-coated glass slide using your finger pad while wearing a disposable latex glove. The tape should be smaller than the glass and should completely cover the sample leaf (Fig. 1a). Be careful to avoid creating air bubbles between the tape and the glass.
2. Detach the inoculated leaf and place it onto the double-sided tape fixed to the slide glass prepared above (Fig. 1a).
3. Press the leaf firmly, but gently, against the tape using your finger pad while wearing a disposable latex glove.
4. Set the leaf specimen on the slide glass directly under the objective lens of the stereomicroscope for imaging. Acquire a YFP image of the entire leaf sample using a fluorescence stereo microscope. Acquire an image of chlorophyll autofluorescence using the TXR filter, as well as a normal bright field image, to identify the leaf shape and the inoculated leaf area undergoing hypersensitive cell death, as reflected by loss of chlorophyll autofluorescence (Fig. 1b).

3.4 MALDI-MSI

The imaging mass microscope iMScope QT is an MS imaging system with atmospheric MALDI equipped with an optical microscope combined with an LCMS Q-TOF. To perform MALDI-MSI, it is necessary to uniformly apply a matrix, which acts as an ionization enhancer, prior to MS analysis. A fresh leaf sample on the taped ITO glass (Subheading 3.4, **step 4**) can be directly used for matrix deposition as described below.

1. Use CHCA as the matrix for ionization of CA, with the aid of iMLayer, which enables uniform deposition of the matrix on the sample, leading to higher reproducibility (*see* **Note 4**).
2. Place CHCA and the prepared ITO glass (Subheading 3.4, **step 4**) appropriately into the iMLayer instrument, following the manufacturer's instructions.
3. Set the thickness of the matrix to 0.7 μm and run the iMLayer deposition program.
4. After the deposition program is complete, remove the CHCA-coated ITO glass and fit it into the sample holder compatible with iMScope QT.
5. Insert the sample holder carrying the CHCA-coated ITO glass into the iMScope QT instrument.

Table 1
MALDI-MSI analysis parameters

Mass spectrometer	
System	iMScope QT + LCMS-9050
Polarity	Positive
Mass range	*m/z* 200–210
Spatial resolution (pitch)	50 μm
Laser diameter setting	4
Laser intensity	60
Laser repetition frequency	1 kHz
Matrix coating	
System	iMLayer
Matrix used	α-Cyano-4-hydroxycinnamic acid (CHCA)
Coating method	Deposition with 0.7 μm thickness

6. Capture a bright field image of the leaf sample using the CCD camera of iMScope QT and set the focus to the surface of the leaf.
7. Define the MSI area to fully cover the sample leaf.
8. Set the parameters accordingly and run the MSI scanning program (Table 1).
9. The data is automatically generated (Fig. 2b–d).

3.5 LA-ICP-MSI

The sample on the CHCA-coated ITO glass that was analyzed using iMScope QT can be sequentially analyzed using ICPMS-2030 coupled with imageBIO266.

1. Remove the sample holder carrying the CHCA-coated ITO glass from the iMScope QT instrument.
2. Set the CHCA-coated ITO glass on a sample holder of imageBIO266.
3. Insert the sample holder carrying the CHCA-coated ITO glass into imageBIO266.
4. Set the focus to the surface of the leaf and define an analysis area by referring to the laser irradiation mark of iMScope QT.
5. Set the parameters according to the manufacturer's instructions (Table 2) and run the LA-ICP-MS measurement.
6. The data is automatically generated (Fig. 2e).

Table 2
LA-ICP-MSI analysis parameters

ICP-MS	
System	ICPMS-2030
Radio frequency power	1.20 kW
Sampling depth	5.0 mm
Plasma gas	9.0 L min^{-1}
Auxiliary gas	0.70 L min^{-1}
Carrier gas	0.40 L min^{-1}
Chamber temperature	5 °C
Peristaltic pump rotation speed (rpm)	20 rpm
LA	
System	imageBIO266
Laser energy	11%
Laser fluence	0.8 J cm^{-2}
Laser spot diameter	50 μm
Scan speed	100 μm s^{-1}
Laser repetition rate	500 Hz
Helium gas flow	0.800 L min^{-1}

3.6 MSI Data Analysis (Visualization)

MALDI-MSI data and LA-ICP-MSI data can be analyzed using a single software. The generation and comparison of MS images are performed using the MS imaging analysis software IMAGEREVEAL MS.

4 Notes

1. It is advisable to use a Petri dish of sufficient height that does not interfere with the plants. An aseptic growth condition is recommended to avoid unexpected activation of CA synthesis by uncontrolled environmental factors.
2. Unpublished material. Briefly, the 3.0-kb upstream sequence was used as the *PAD3* promoter and fused to eYFP-NLS using the pBGYN vector, as reported for the construction of pPR1-eYFP-NLS and pVSP1-eYFP-NLS plants [11, 12].
3. Partial, not full, infiltration of the leaf is recommended to visualize local CA accumulation only around the infiltrated area. Avoid dropping the bacterial suspension onto the media (especially, the root) and the other part of the plant during

infiltration, which will trigger immune responses in non-infiltrated parts of the plant.

4. When analyzing a new metabolite, an appropriate matrix must be evaluated and selected using the standard of the metabolite in the same system.

Acknowledgments

We would like to thank Ms. Eriko Betsuyaku for supporting the maintenance of the plants. This work was supported by the Japan Society for the Promotion of Science (JSPS) KAKENHI, Grant-in-Aid for Transformative Research Areas (B) (23H03847), the Japan Science and Technology Agency PRESTO (117665 to SB), and ERATO (JPMJER1502 to SB).

References

1. Ahuja I, Kissen R, Bones AM (2012) Phytoalexins in defense against pathogens. Trends Plant Sci 17:73–90
2. Ebel J (1986) Phytoalexin synthesis: the biochemical analysis of the induction process. Annu Rev Phytopathol 24:235–264
3. Somssich IE, Hahlbrock K (1998) Pathogen defence in plants — a paradigm of biological complexity. Trends Plant Sci 3:86–90
4. Schmelzer E, Kruger-Lebus S, Hahlbrock K (1989) Temporal and spatial patterns of gene expression around sites of attempted fungal infection in parsley leaves. Plant Cell 1:993–1001
5. Schuhegger R, Nafisi M, Mansourova M, Petersen BL, Olsen CE, Svatoš A, Halkier BA, Glawischnig E (2006) CYP71B15 (PAD3) catalyzes the final step in camalexin biosynthesis. Plant Physiol 141:1248–1254
6. Snyder BA, Nicholson RL (1990) Synthesis of phytoalexins in sorghum as a site-specific response to fungal ingress. Science 248:1637–1639
7. Boughton BA, Thinagaran D, Sarabia D, Bacic A, Roessner U (2016) Mass spectrometry imaging for plant biology: a review. Phytochem Rev 15:445–488
8. Ryffel F, Helfrich EJ, Kiefer P, Peyriga L, Portais J-C, Piel J, Vorholt JA (2016) Metabolic footprint of epiphytic bacteria on Arabidopsis thaliana leaves. ISME J 10:632–643
9. Glazebrook J, Ausubel FM (1994) Isolation of phytoalexin-deficient mutants of Arabidopsis thaliana and characterization of their interactions with bacterial pathogens. Proc Natl Acad Sci 91:8955–8959
10. Zhou N, Tootle TL, Glazebrook J (1999) Arabidopsis PAD3, a gene required for camalexin biosynthesis, encodes a putative cytochrome P450 monooxygenase. Plant Cell 11:2419–2428
11. Betsuyaku S, Katou S, Takebayashi Y, Sakakibara H, Nomura N, Fukuda H (2018) Salicylic acid and jasmonic acid pathways are activated in spatially different domains around the infection site during effector-triggered immunity in *Arabidopsis thaliana*. Plant Cell Physiol 59:439–439
12. Kubo M, Udagawa M, Nishikubo N, Horiguchi G, Yamaguchi M, Ito J, Mimura T, Fukuda H, Demura T (2005) Transcription switches for protoxylem and metaxylem vessel formation. Genes Dev 19:1855–1860

Chapter 16

Bioluminescence-Based Growth Quantification of the Phytopathogenic Bacterium *Pseudomonas syringae* pv. *tomato* DC3000

Yuga Fujinawa, Yijia Yan, Hirofumi Nakagami, and Akira Mine

Abstract

Bioluminescence tagging has gained popularity as an effective tool for investigating infection processes of phytopathogenic bacteria. A critical consideration in employing this approach is to minimize the impact of the genetically introduced luciferase genes on bacterial fitness, while maximizing the intensity and stability of bioluminescence. Recently, the pBJ vector series was developed as all-in-one system for bioluminescence tagging in a wide range of Pseudomonadota, including the phytopathogenic bacterium *Pseudomonas syringae* pv. *tomato* DC3000 (*Pto*). The pBJ vectors enable inducible transposition of the *luxCDABE* luciferase operon, derived from *Photorhabdus luminescens*, into a specific and neutral genomic location via Tn*7* transposon. The resulting bioluminescent *Pto* strain, termed *Pto*-lux, emits stable and strong bioluminescence while maintaining bacterial fitness during plant infection. Moreover, bioluminescence-based assays using *Pto*-lux enable accurate quantification of *in planta* bacterial titers across diverse host plants, with a dynamic range of four orders of magnitude. In this chapter, we present detailed protocols for the bioluminescence-based bacterial growth assays in *Arabidopsis thaliana* and *Marchantia polymorpha*.

Key words Plant-bacteria interactions, *Pseudomonas syringae*, *Arabidopsis thaliana*, *Marchantia polymorpha*, Bioluminescence, *luxCDABE*, Plant immunity, Bacterial virulence

1 Introduction

The phytopathogenic bacterium *Pseudomonas syringae* pv. *tomato* DC3000 (*Pto*) was originally isolated from tomato but was later found to infect the model angiosperm *Arabidopsis thaliana* and the model liverwort *Marchantia polymorpha*. Owing to its broad host range, *Pto* has served as a model organism for studying the molecular and evolutionary aspects of plant-bacterial pathogen interactions [1, 2].

Accurate quantification of *in planta* bacterial titers is essential for investigating mechanisms of plant immunity and bacterial virulence. Although conventional colony counting assays remain a

Tsutomu Kawasaki and Hirofumi Yoshioka (eds.), *Plant Immunity: Methods and Protocols*, Methods in Molecular Biology, vol. 3012, https://doi.org/10.1007/978-1-0716-5138-4_16,

standard method for quantifying bacterial growth in infected plants, bioluminescence-based assays have gained increasing popularity [3–6]. The generation of bioluminescent bacteria typically employs the *luxCDABE* luciferase operon of *Photorhabdus luminescens*, which enables metabolically active bacterial cells to emit light at a wavelength of 490 nm without an external supply of luciferase substrates [7]. A critical consideration in generating bioluminescent bacteria is to ensure that the introduction of *luxCDABE* does not compromise bacterial fitness, while still achieving robust and stable light emission. Recently, pBJ vectors were developed as a versatile system for Tn*7* transposon-mediated integration of *luxCDABE* into the neutral attTn*7* site in the genomes of various Pseudomonadota (Proteobacteria) [3]. Among them, the pBJ2 vector was used to generate a bioluminescent strain of *Pto*, termed *Pto*-lux, in which *luxCDABE* is constitutively expressed under the control of kanamycin promoter (also known as *neomycin phosphotransferase II* promoter). *Pto*-lux exhibits strong and stable bioluminescence while maintaining pathogenic fitness during plant infection. Furthermore, bioluminescence emitted by *Pto*-lux serves as an accurate measure to estimate bacterial titers in *A. thaliana*, *M. polymorpha*, *Solanum lycopersicum*, and *Nicotiana benthamiana*, with a dynamic range of four orders of magnitude [3].

In this chapter, we describe detailed protocols for bioluminescence-based growth assays using *Pto*-lux. Using these methods, we demonstrate reliable quantification of the enhanced susceptibility of the *A. thaliana sid2–2* mutant, which is deficient in the production of the defense hormone salicylic acid [8]. We also highlight the practical application of this assay in *M. polymorpha*, an emerging model system for elucidating evolutionary aspects of plant-bacterial pathogen interactions [9, 10].

2 Materials

2.1 Preparation of Pto-*lux* Inoculum

1. *Pto*-lux: *Pseudomonas syringae* pv. *tomato* DC3000 chromosomally tagged with the kanamycin promoter-*luxCDABE* cassette at the attTn*7* site.
2. King's B (KB) medium: Dissolve 4 g of tryptone, 0.3 g of dipotassium phosphate, 3 g of glycerol in 200 mL of pure water. Adjust the pH to 7.2 using 1 M KOH. Autoclave at 121 °C for 15 min (*see* **Note 1**).
3. Agar.
4. Rifampicin.
5. Petri dish.
6. Shaking incubator.

2.2 Syringe Infiltration into A. thaliana

1. *A. thaliana* ecotype Columbia (Col-0) and *sid2–2* mutant.
2. Plant growth chamber.
3. Centrifuge.
4. Spectrophotometer.
5. Needleless syringe.

2.3 Vacuum Infiltration into M. polymorpha

1. *M. polymorpha* accession Tak-1.
2. Gamborg's B5 medium (half-strength).
3. Plant agar.
4. Petri dish (90 mm diameter × 25 mm height).
5. Whatman™ filter paper (Grade 1, diameter 85 mm).
6. Aluminum foil.
7. Sterile tweezer.
8. Plant growth chamber.
9. Spectrophotometer.
10. Centrifuge.
11. Desiccator.
12. Square filter paper.
13. Square Petri dish.
14. Parafilm.
15. 3 M™ Micropore™ Surgical Tape.

2.4 Bioluminescence Measurement

1. Tweezer.
2. Biopsy punch.
3. White, light-reflecting 96-well plate (*see* **Note 2**).
4. Microplate luminometer.

3 Methods

3.1 Measurement of Pto-*lux* Growth in A. thaliana

3.1.1 Growing Plants

1. To break dormancy, incubate *A. thaliana seeds* in deionized water at 4 °C for 4 days in the dark.
2. Sow the seeds in plastic pots filled with soil (*see* **Note 3**).
3. Place the pots in a plant growth chamber equipped with white fluorescent light. Maintain the following growth conditions: temperature of 22 °C, light intensity of 6000 lux (ca. 100 μmol/m^2/s) for 10 h, and 60% relative humidity.
4. Water plants with deionized water containing a 2000-fold dilution of Hyponex fertilizer when the soil surface appears dry.

3.1.2 Pto-*lux Preparation*

1. Prepare KB medium solidified with 1.5% agar and supplemented with 50 μg/mL rifampicin, and pour into Petri dishes (*see* **Note 4**).
2. Streak *Pto*-lux onto solid KB medium.
3. Incubate for 2–3 days at 28 °C.
4. Suspend several single colonies of *Pto*-lux into 200 μL of liquid KB medium containing 50 μg/mL rifampicin (*see* **Note 5**).
5. Transfer the suspension into 4 mL of liquid KB medium containing 50 μg/mL rifampicin in a test tube (*see* **Note 6**).
6. Incubate the liquid culture overnight at 28 °C with shaking at 200 rpm.

3.1.3 *Infiltration into* A. thaliana *Leaves*

1. Transfer the overnight culture in the logarithmic growth phase into a 2 mL centrifugation tube (*see* **Note 6**).
2. Centrifuge the culture at 8000 rpm for 2 min at room temperature.
3. Decant the supernatant and resuspend the pellet in 1 mL of autoclaved pure water.
4. Centrifuge the suspension at 8000 rpm for 2 min, remove the supernatant, and resuspend the pellet in 1 mL of autoclaved pure water.
5. Repeat step 4 once more.
6. Measure the OD_{600} of the *Pto*-lux suspension (*see* **Note 7**).
7. Adjust the suspension to OD_{600} of 0.0005 using autoclaved pure water (*see* **Note 8**).
8. Infiltrate the bacterial suspension using a 1 mL needleless syringe into three leaves of 4- to 5-week-old *A. thaliana* (*see* **Note 9**).
9. Incubate the inoculated plants for 2 days post-inoculation (dpi) at 22 °C in a growth chamber with the following conditions: temperature of 22 °C, light intensity of 6000 lux (ca. 100 μmol/m^2/s) for 10 h, and relative humidity of 60%.

3.1.4 *Bioluminescence Measurement*

1. Detach the infiltrated leaves from the petiole using a tweezer.
2. Excise four leaf discs per leaf using a 4 mm biopsy punch.
3. Place leaf discs into individual wells of a white 96-well microplate (Fig. 1).
4. Set the microplate in a GloMax Navigator Microplate Luminometer (Promega), close the lid, and incubate for 10 min in the dark (*see* **Note 10**).
5. Measure bioluminescence from each well with an acquisition time of 10 s (*see* **Note 11**).

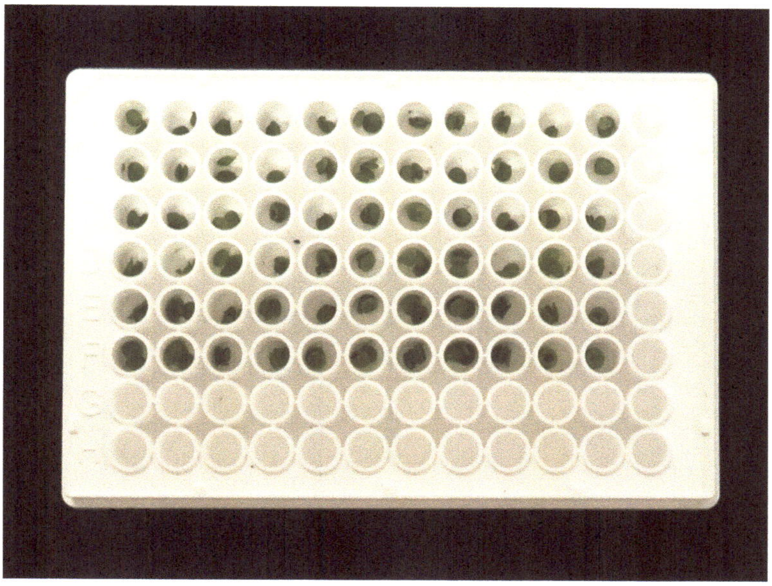

Fig. 1 A 96-well plate used for bioluminescence measurement. Four leaf discs were excised from a single *A. thaliana* leaf infiltrated with *Pto*-lux. Leaf discs from different leaves were placed into individual wells of a 96-well plate for bioluminescence detection

3.1.5 Colony Counting (Optional)

1. After bioluminescence measurement, transfer leaf discs into an 8-strip collection tube containing two 4 mm stainless steel beads.
2. Add 400 μL of 5 mM $MgSO_4$ and pulverize the tissue by bead beating.
3. Make serial dilutions up to 10^6.
4. Streak 10 μL from each dilution on KB agar containing 50 μg/mL rifampicin.
5. Incubate the plates at 28 °C for 2–3 days, then count colonies (*see* **Note 12**).

3.1.6 Data Analysis (Optional)

1. Calculate the $\log_{10}$-transformed relative light units (RLUs) and colony-forming units (CFUs) per cm^2 leaf surface area.
2. Calculate and visualize the Pearson correlation coefficient between RLUs and CFUs (Fig. 2) (*see* **Note 13**).
3. Visualize RLUs using boxplots (Fig. 3) (*see* **Note 14**).

3.2 Measurement of Pto-*lux* Growth in M. polymorpha

3.2.1 Growing Plants

1. Wrap Whatman™ filter papers in aluminum foil and autoclave them once in a heat-stable container prior to use.
2. Prepare half-strength Gamborg's B5 medium with 1% plant agar and pour into Petri dishes (90 mm diameter × 25 mm height).

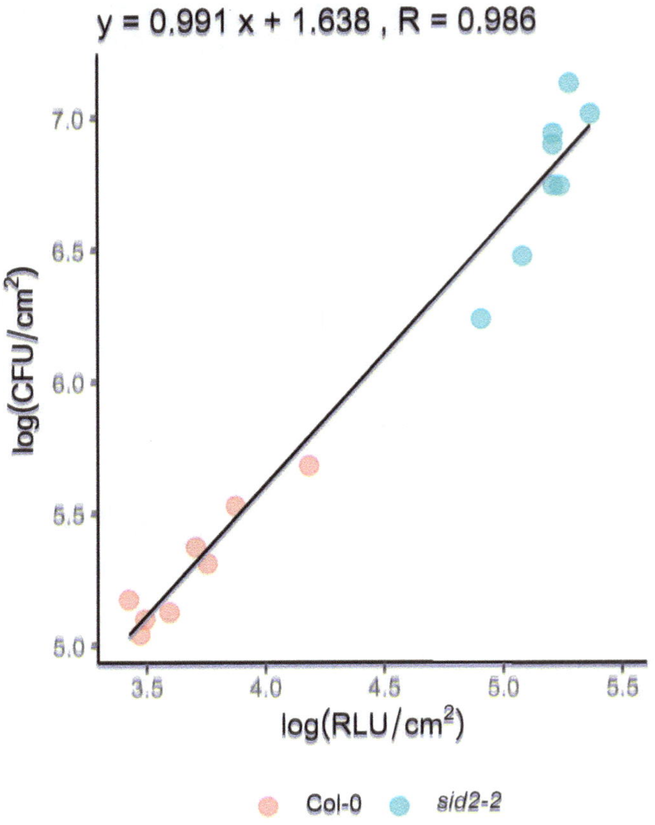

Fig. 2 High correlation between bacterial titers estimated by colony counting assays and bioluminescence-based assays. Leaves of *A. thaliana* wild-type Col-0 and *sid2-2* mutant were infiltrated with *Pto*-lux at OD_{600} of 0.0005. Two days after infiltration, bioluminescence (relative light units: RLUs) was determined, followed by quantification of bacterial titers as colony-forming units (CFUs). Log_{10}-transformed RLUs and CFUs per leaf area (cm^2) were calculated and plotted along the x and y axes, respectively. Regression equations and Pearson's correlation coefficient (*R*) are shown above the plot

3. Place a single layer of Whatman™ filter paper on top of the solid medium using a sterile tweezer (*see* **Note 15**).
4. Transfer individual *M. polymorpha* gemmae onto the filter paper on the solid medium (*see* **Note 16**).
5. Grow *M. polymorpha* gemmae for 14 days in a growth chamber under constant white LED light (50–60 μmol/m^2/s) at 22–24 °C.

3.2.2 *Pto-lux* Preparation

1. Prepare KB medium solidified with 1.5% agar and supplemented with 25 μg/mL rifampicin and pour into Petri dishes (*see* **Note 4**).
2. Streak *Pto*-lux onto solid KB medium.

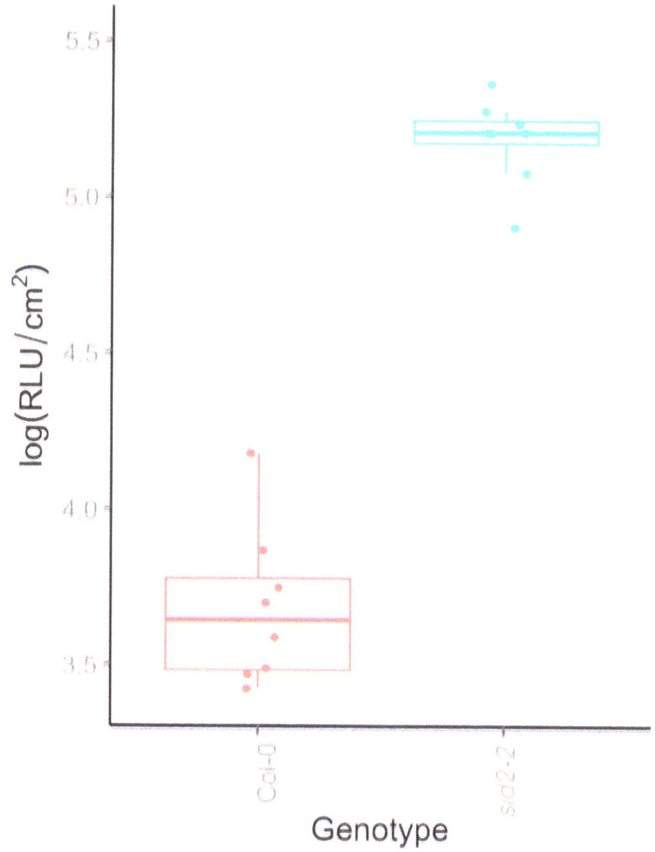

Fig. 3 Bioluminescence-based growth assay detects the enhanced susceptibility of *sid2-2* compared to Col-0. Leaves of *A. thaliana* wild-type Col-0 and *sid2-2* mutant were infiltrated with *Pto*-lux at OD_{600} of 0.0005. Two days after infiltration, bacterial titers were determined as RLUs. Log_{10}-transformed RLUs were calculated and plotted. A statistically significant difference between Col-0 and *sid2-2* was detected ($P = 6.5 \times 10^{-10}$, Student's t-test)

3. Incubate overnight at room temperature.
4. Inoculate a single colony of *Pto*-lux into 10 mL of liquid KB medium containing 25 μg/mL rifampicin in a 50 mL falcon tube.
5. Incubate the liquid culture overnight at 28 °C with shaking at 200 rpm (*see* **Note 17**).

3.2.3 Vacuum Infiltration of Pto-lux

1. Wrap square filter papers in aluminum foil and autoclave them once in a heat-stable container prior to use.
2. Place an autoclaved square filter paper in each square plate.
3. Moisten the filter paper with 1 mL autoclaved ultrapure water using a pipette.

4. Transfer individual *M. polymorpha* thalli into an empty Petri dish (90 mm diameter × 25 mm height) using sterile tweezer (*see* **Note 18**).
5. Measure the OD_{600} of *Pto*-lux culture and harvest when it reaches 0.8 (*see* **Note 19**).
6. Centrifuge the culture at 4000 rpm for 10 min at 4 °C.
7. Decant the supernatant and resuspend the pellet in autoclaved ultrapure water.
8. Adjust the suspension to OD_{600} of 0.01 using autoclaved ultrapure water.
9. Submerge the thalli in the bacterial suspension within a Petri dish (90 mm diameter × 25 mm height) and vacuum infiltrate using a desiccator for 5 min (*see* **Note 20**).
10. Decant the suspension and transfer the infiltrated plants onto the moistened filter paper in the square Petri dishes, typically placing 8 plants per plate.
11. Add 1 mL of autoclaved ultrapure water to each filter paper using a pipette (*see* **Note 21**).
12. Seal the plates with micropore tape.
13. Incubate for 2 days post-inoculation (dpi) at 22 °C in a growth chamber under 50 μmol/m^2/s light intensity with a 16 h light/8 h dark cycle (*see* **Note 22**).

3.2.4 Bioluminescence Measurement

1. Excise one apical and one basal thallus disc from each plant using a 5 mm biopsy punch (*see* **Note 23**).
2. Place samples rhizoids-down into a white 96-well microplate.
3. Measure bioluminescence using the preset measurement protocol "LUMI-TEST" in the FLUOstar Omega microplate reader (BMG LABTECH) (*see* **Note 10**).

4 Notes

1. Although the original KB medium recipe includes $MgSO_4$, *Pto*-lux can grow well in KB medium without it.
2. We use white 96-well plates from Corning (Cat. No. 3912) and Greiner Bio-One (Cat. No. 655074) for bioluminescence-based assays in *A. thaliana* and *M. polymorpha*, respectively; however, any white light-reflecting 96 plates should work.
3. We use a 1:1 mixture of Supermix A (Sakataseed) and vermiculite G20 (Nittai); however, any available soil that supports healthy growth of *A. thaliana* is acceptable.

4. *Pto*-lux can be selectively cultured using rifampicin at 25–50 μg/mL.
5. It is recommended to resuspend 5–10 single colonies to ensure robust and rapid growth of *Pto*-lux.
6. Prepare serial dilutions of the inoculum to ensure that at least one overnight culture reaches the logarithmic growth phase.
7. It is recommended to prepare a 1:10 dilution of bacterial suspension to ensure that the OD_{600} value does not exceed the upper quantification limit.
8. The dynamic range of the bioluminescence-based growth assay reliably encompasses bacterial populations derived from inocula with OD_{600} values between 0.0002 and 0.01 [3].
9. 6th, 7th and 8th leaves are preferable for infiltration. 5th and 9th leaves can also be used.
10. To minimize background from chlorophyll fluorescence, acclimate the samples in the dark within the plate reader for 10 min prior to measurement. Bioluminescence imaging can alternatively be performed using the ChemiDoc XRS+ system with an appropriate signal accumulation protocol.
11. 96-well plates can be reused after disinfection by soaking in 70% ethanol for more than 30 min. Ensure the plates are completely dry before use, and confirm that none of the wells show luminescence levels higher than background (~10–20 RLU).
12. Avoid using plates with too few or too many colonies. A reliable range for colony counting is 10–100 colonies.
13. The R function "cor" can be used to calculate Pearson correlation coefficients.
14. The R package "ggplot2" can be used to make a boxplot.
15. The filter paper becomes saturated with medium; carefully remove air bubbles between the filter paper and agar medium using a sterile tweezer.
16. Limit to a maximum of 8 gemmae per Petri dish to allow sufficient space for thallus growth.
17. The required volume of liquid culture may vary depending on experimental scale and inoculation method. Slightly loosen the falcon tube cap and seal with micropore tape to ensure adequate aeration.
18. Carefully remove the filter paper from the thalli without damaging them. If the filter paper adheres to the rhizoids and is difficult to remove, leave it attached. Collect thalli of the same genotype in an empty Petri dish and close the lid to avoid desiccation.

19. Cultures with OD_{600} between 0.8 and 1.0 are acceptable for inoculation [3].
20. Apply vacuum infiltration until no air bubbles are visible from the thalli. Do not exceed 5 min to prevent damage to the tissue.
21. Ensure that the water does not come into contact with the surface of the thalli.
22. At 1 dpi, replace micropore tape with parafilm to reduce desiccation. For longer incubations (>3 dpi), supplement the filter paper with additional autoclaved ultrapure water as needed.
23. The diameter of the biopsy punch may vary depending on the experimental design and the size of the thalli. For 14-day-old wild-type *M. polymorpha* thalli, a diameter of 4–5 mm is generally suitable. Excise plant discs on a bench, and sterilize the tip of the biopsy punch by briefly flaming it before sampling and when switching between different genotypes to prevent microbial cross-contamination.

Acknowledgments

This work was supported by Grant-in-Aid for Scientific Research (B) (24 K01757), Grant-in-Aid for Challenging Exploratory Research (22 K19178), and JST GteX (JPMJGX23B2). This work was also supported by the Max Planck Society and was conducted in the framework of MAdLand (http://madland.science, Deutsche Forschungsgemeinschaft [DFG] Priority Programme 2237). H.N. is grateful for funding by the DFG (NA 946/1-1 and NA 946/1-2). Y.Y. is grateful to the China Scholarship Council (CSC) for PhD studentship funding.

References

1. Gimenez-Ibanez S, Zamarreno AM, Garcia-Mina JM, Solano R (2019) An evolutionarily ancient immune system governs the interactions between pseudomonas syringae and an early-diverging land plant lineage. Curr Biol 29(14):2270–2281 e2274. https://doi.org/10.1016/j.cub.2019.05.079
2. Xin XF, Kvitko B, He SY (2018) Pseudomonas syringae: what it takes to be a pathogen. Nat Rev Microbiol 16(5):316–328. https://doi.org/10.1038/nrmicro.2018.17
3. Matsumoto A, Schluter T, Melkonian K, Takeda A, Nakagami H, Mine A (2022) A versatile Tn7 transposon-based bioluminescence tagging tool for quantitative and spatial detection of bacteria in plants. Plant Commun 3(1):100227. https://doi.org/10.1016/j.xplc.2021.100227
4. Taks NW, Batstra MD, Kortekaas RF, Stevens FD, Pfeilmeier S, van den Burg HA (2025) Non-invasive, bioluminescence-based visualisation and quantification of bacterial infections in arabidopsis over time. Mol Plant Pathol 26(2):e70055. https://doi.org/10.1111/mpp.70055
5. Soldan R, Sanguankiattichai N, Bach-Pages M, Bervoets I, Huang WE, Preston GM (2021) From macro to micro: a combined bioluminescence-fluorescence approach to monitor bacterial localization. Environ Microbiol 23(4):2070–2085. https://doi.org/10.1111/1462-2920.15296

6. Li Q, Zhou M, Chhajed S, Yu F, Chen S, Zhang Y, Mou Z (2023) N-hydroxypipecolic acid triggers systemic acquired resistance through extracellular NAD(P). Nat Commun 14(1):6848. https://doi.org/10.1038/s41467-023-42629-0
7. Meighen EA (1993) Bacterial bioluminescence: organization, regulation, and application of the lux genes. FASEB J 7:1016–1022
8. Wildermuth MC, Dewdney J, Wu G, Ausubel FM (2001) Isochorismate synthase is required to synthesize salicylic acid for plant defence. Nature 414(6863):562–565. https://doi.org/10.1038/35107108
9. Jeon HW, Iwakawa H, Naramoto S, Herrfurth C, Gutsche N, Schluter T, Kyozuka J, Miyauchi S, Feussner I, Zachgo S, Nakagami H (2024) Contrasting and conserved roles of NPR pathways in diverged land plant lineages. New Phytol 243(6): 2295–2310. https://doi.org/10.1111/nph.19981
10. Yotsui I, Matsui H, Miyauchi S, Iwakawa H, Melkonian K, Schluter T, Michavila S, Kanazawa T, Nomura Y, Stolze SC, Jeon HW, Yan Y, Harzen A, Sugano SS, Shirakawa M, Nishihama R, Ichihashi Y, Ibanez SG, Shirasu K, Ueda T, Kohchi T, Nakagami H (2023) LysM-mediated signaling in Marchantia polymorpha highlights the conservation of pattern-triggered immunity in land plants. Curr Biol 33(17):3732–3746 e3738. https://doi.org/10.1016/j.cub.2023.07.068

Chapter 17

Genome Editing for Developing Disease-Resistant Plants

Muhammad Sohaib Shafique, Yapei Liu, and Zhiyuan Ji

Abstract

Genome editing tools, most notably the CRISPR-Cas9 system, have transformed modern crop breeding by allowing precise and efficient modification of plant genomes, particularly for enhancing resistance to disease. In this chapter, we present two major CRISPR-based approaches that have been successfully applied to develop disease-resistant crops: (1) gene knockout of host susceptibility (*S*) genes to block pathogen compatibility, and (2) targeted insertion of resistance alleles or synthetic genes using the homology-directed repair (HDR) pathway. Both strategies utilize the specificity and efficiency of CRISPR-Cas9 to introduce defined edits in crops such as rice. This chapter outlines comprehensive step-by-step protocols covering key elements of the process, including target site selection, construction of sgRNA and Cas9 vectors, plant transformation methods, and mutation validation techniques. Together, these methods provide a powerful platform for engineering plant immune responses and offer practical tools to accelerate the breeding of disease-resistant cultivars.

Key words CRISPR-Cas9, Gene knockout, Targeted knock-in, Disease resistance, Genome editing

1 Introduction

Plant diseases caused by bacterial, fungal, and viral pathogens significantly threaten global food security, causing severe yield losses, particularly affecting staple crops essential for caloric nutrition [1]. Bacterial pathogens, for example, *Xanthomonas oryzae* (causing bacterial blight in rice) and *Pseudomonas syringae* (causing bacterial speck in tomatoes), exploit host susceptibility genes for infection [2]. Genetic resistance is durable and sustainable approach to combat plant diseases. In contrast, fungal pathogen *Magnaporthe oryzae* causing rice blast and *Puccinia graminis* causing wheat stem rust rapidly evolve to overcome conventional genetic resistance. Viral pathogens including Tomato yellow leaf curl virus and Cassava mosaic virus continue to cause devastating epidemics in staple crops worldwide [3]. Traditional plant breeding methods for developing disease-resistant cultivars are very effective but are hindered by their dependence on naturally occurring

Tsutomu Kawasaki and Hirofumi Yoshioka (eds.), *Plant Immunity: Methods and Protocols*, Methods in Molecular Biology, vol. 3012,
https://doi.org/10.1007/978-1-0716-5138-4_17,

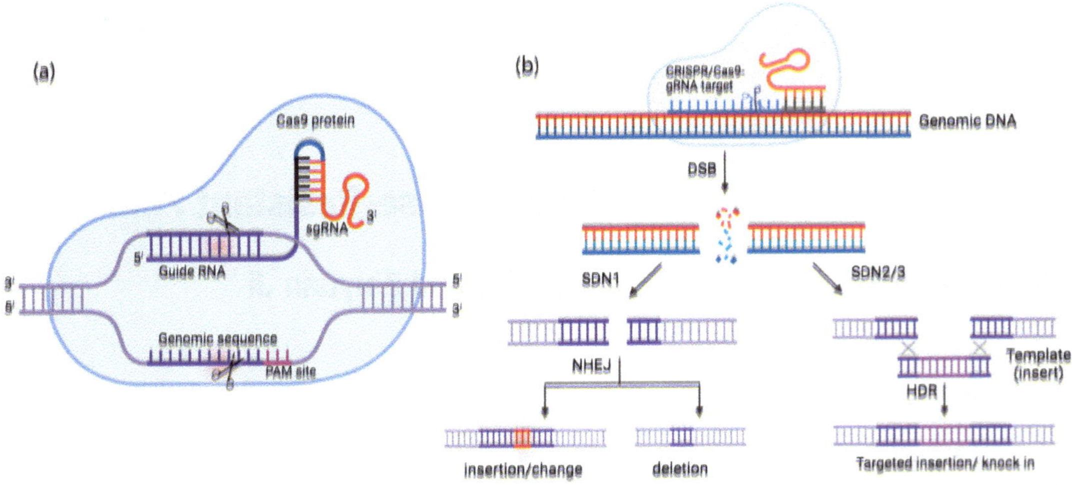

Fig. 1 CRISPR/Cas9-mediated genome editing mechanism and repair outcomes. (**a**) Overview of the CRISPR/Cas9 system. The Cas9 endonuclease is guided by an sgRNA to introduce a double-strand break (DSB) at a specific genomic locus adjacent to a protospacer adjacent motif (PAM). (**b**) The DSB can be repaired by two major pathways: NHEJ, which often results in insertions or deletions (indels) leading to gene disruption; or HDR, which uses a donor template to enable precise modifications or targeted gene insertion

genetic variation and typically require extensive backcrossing over 5–10 years to eliminate undesirable linked traits. Moreover, the durability of resistance achieved through traditional breeding is often not prolonged, as rapidly evolving pathogen populations are able to overcome it over time.

Clustered Regularly Interspaced Short Palindromic Repeats (CRISPR)-Cas9 is a programmable RNA-guided endonuclease system adapted from bacterial immunity and widely used for targeted genome editing. It consists of a single guide RNA (sgRNA) that directs the Cas9 protein to a specific DNA sequence, where Cas9 induces a double-strand break (DSB) (Fig. 1a). The DSB is repaired by the cell through non-homologous end joining (NHEJ) or homology-directed repair (HDR) (Fig. 1b). NHEJ often results in insertions or deletions (indels) at the break site, referred to as site-directed nuclease 1 (SDN1). When a repair template is supplied, HDR enables precise modifications (SDN2) or the insertion of larger sequences (SDN3). Hence, the potential of SDN1, SDN2, and SDN3 has been widely utilized in *CRISPR*-based gene knockout and targeted insertion strategies to develop disease-resistant plants.

In recent years, CRISPR-Cas9 genome editing has revolutionized the development of disease-resistant crops by enabling precise and efficient genetic modifications. Through the targeted disruption of susceptibility genes, such as *OsSWEET14* in rice and *MLO* in wheat, researchers have been able to generate plants with broad-spectrum resistance to bacterial and fungal pathogens, respectively [4, 5]. Beyond straightforward gene knockouts, CRISPR technologies have expanded to enable precise modifications within critical effector-binding sites of host proteins, directly disrupting pathogen virulence strategies without compromising plant growth or fitness [6]. For example, precise single-amino-acid substitutions in the rice transcription factor *OsERF922* significantly enhance resistance to rice blast caused by *Magnaporthe oryzae* [7]. Additionally, advanced genome editing approaches employing homology-directed repair (HDR) have enabled targeted insertion (knock-in) of desired sequences of choice [6, 8] or engineered immune receptors [9], substantially expanding available genetic resources beyond natural allelic diversity.

This chapter provides comprehensive and practical genome editing protocols detailing two primary CRISPR-Cas9 strategies: gene knockout and targeted insertion. The protocols described herein are specifically optimized for rice but can be broadly applied to other monocot and dicot species, with an emphasis on essential procedures such as guide RNA selection, vector assembly, *Agrobacterium*-mediated transformation, and rigorous mutation validation. The integration of these genome editing tools into crop improvement programs can accelerate the generation of durable, disease-resistant cultivars, presenting a sustainable solution to mitigate yield losses and support global food production. Highlighted case studies illustrate the successful deployment of genome editing approaches, including targeted disruption of the rice *OsSULTR3;6* gene to confer resistance against bacterial leaf streak [10, 11] and targeted insertion of the bacterial blight resistance gene *Xa23* in rice [6, 12]. With ongoing advancements in precision editing methodologies and delivery systems, genome editing technologies hold tremendous promise for revolutionizing crop protection strategies, ultimately enhancing resilience against evolving pathogen threats.

2 Materials

Prepare all reagents using high-quality, deionized water and analytical-grade chemicals. Keep restriction enzymes on ice to preserve their activity and avoid repeated freeze-thaw cycles. Maintain sterile conditions when preparing vectors, primers, and other DNA components. For cloning and transformation steps, always use nuclease-free materials to prevent contamination. Antibiotics should be added to the media just before use, and all bacterial cultures should be handled in accordance with standard biosafety protocols. Properly dispose of any genetically modified materials, following the lab's safety guidelines.

2.1 CRISPR/Cas9-Based Genome Editing for Targeted Gene Knockout

1. *E. coli* strains: DH5α or DH10B or TOP10F′.
2. *Agrobacterium tumefaciens*: EHA105.
3. Plasmids: Intermediate sgRNA vector pENTR-sgRNA and Cas9 vector pDEST-Cas9.
4. Restriction enzymes BtgZI, BsaI, ApaI, and AatII.
5. T4 DNA Ligase.
6. Gateway® LR Clonase™ II Enzyme Mix (Invitrogen).
7. 10 mM ATP.
8. DNA polymerase: Phanta Max or KOD Plus.
9. Antibiotics: Kanamycin, spectinomycin, hygromycin.
10. Induction plate for boiling water.
11. PCR thermocycler.
12. Gel electrophoresis apparatus.
13. Gel extraction kit.
14. Incubator (37 °C, 28 °C).

2.2 CRISPR/Cas9-Mediated Targeted Insertion via Homology-Directed Repair (HDR)

1. CRISPR/Cas9 expression vector: pCxa23 (expressing Cas9 and sgRNA targeting *xa23* promoter).
2. Donor repair vector: pDxa23 (carrying left homology arm–EBE fragment–right homology arm).
3. *E. coli* DH5α or TOP10F′.
4. *Agrobacterium tumefaciens* EHA105.
5. High-fidelity DNA polymerase (e.g., Phanta Max).
6. Restriction enzymes EcoRI, HindIII.
7. Gel extraction and plasmid purification kits.
8. Regeneration medium consists of MS basal salts supplemented with 30 g/L sucrose, 2–3 mg/L 6-benzylaminopurine (BAP), and 50 μg/mL Hygromycin B, solidified with 0.25% gelrite and adjusted to pH 5.8 before autoclaving.

3 Methods

3.1 CRISPR/Cas9-Based Genome Editing for Targeted Gene Knockout

CRISPR/Cas9 enables the precise knockout of plant genes by guiding Cas9 to specific genomic loci using single or multiplexed sgRNAs, typically driven by U6 promoters. These sgRNAs are cloned into an entry vector and assembled with Cas9 into a binary vector using Gateway cloning. This section presents a case study in rice, where the *OsSULTR3;6* gene (*LOC_Os01g52130*), linked to bacterial leaf streak susceptibility, was knocked out using a dual-sgRNA construct (pCas9-SULTR-KO). The construct was introduced into the rice cultivar Nipponbare via *Agrobacterium tumefaciens*-mediated transformation. While the example focuses on rice, the general methodology is applicable across crop species. The following protocol outlines each step from vector construction to plant transformation and mutation validation.

3.1.1 Selection of Target Sequence

1. Select the target gene based on its role in disease susceptibility or resistance. Focus on regions such as early exons (to disrupt translation) or regulatory elements like effector binding elements (EBEs) in the promoter.
2. Design sgRNA target sequences (19–20 bp, $N_{19}NGG$ format) using plant-specific CRISPR design tools. Recommended platforms include the following:
   - TargetDesign: http://skl.scau.edu.cn/targetdesign/
   - CRISPR-P 2.0: http://cbi.hzau.edu.cn/crispr2/
   - CHOPCHOP: https://chopchop.cbu.uib.no/

   These tools automatically assess off-target risk and provide on-target efficiency scores. Avoid GGG PAM sites and ensure targets are confined within single exons.
3. Ensure promoter compatibility: Select guide sequences that begin with G to ensure efficient transcription under U6-type promoters. Finalize 2–4 sgRNAs per gene to allow redundancy and support multiplex editing.

3.1.2 Example of Selection of Target OsSULTR3;6 Susceptibility Gene Editing

1. The *OsSULTR3;6* gene (*LOC_Os01g52130*), associated with bacterial leaf streak susceptibility in rice, was selected for knockout using CRISPR/Cas9.
2. Two critical regions were targeted: the first exon to disrupt gene function. Guide RNAs (19–20 bp, $N_{19}NGG$ format) were designed using TargetDesign (http://skl.scau.edu.cn/targetdesign/), avoiding GGG PAMs and exon-intron boundaries (Fig. 2b) [10].

3.1.3 Constructions of sgRNA Expression Cassettes

The map of vector for pENTR-sgRNA is given in Fig. 2a.

1. Target sgRNAs cloned into the dual-promoter pENTR4-sgRNA entry vectors at either BtgZI or BsaI sites using the following primer formats.
   - For BtgZI cloning:
     - Forward primer: 5′-tgttg-N_{19}–3′.
     - Reverse primer: 5′-aaac-revComp(N_{19})-3′.
   - For BsaI cloning:
     - Forward primer: 5′-gtgtg-N_{19}–3′.
     - Reverse primer: 5′-aaac-revComp(N_{19})-3′.

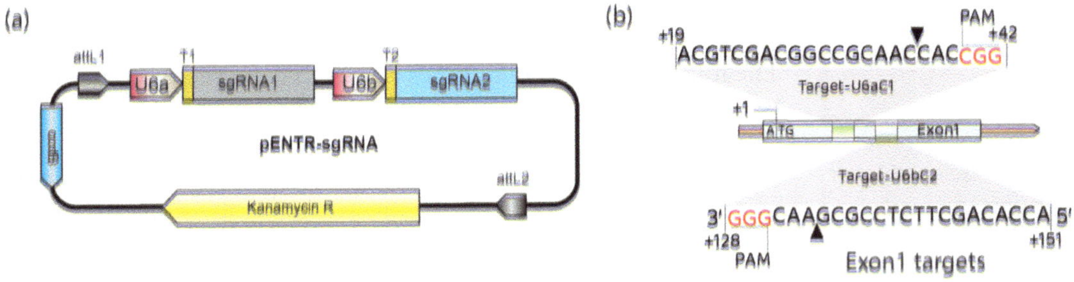

Fig. 2 Dual-sgRNA expression construct and target selection. (**a**) The vector contains two sgRNA expression cassettes driven by U6P1 and U6P2 promoters. sgRNA1 and sgRNA2 are cloned using BtgZI and BsaI restriction sites, respectively. A kanamycin resistance gene enables bacterial selection. (**b**) Two sgRNA target sites (Target-U6aF1 and Target-U6bC2) were designed within exon 1 of *OsSULTR3;6*, flanking the coding region to induce gene knockout. PAM sequences are shown in red, and cleavage sites are indicated by black arrowheads

2. After selecting the target sequences and generating sticky-end primers, boil water to 100 °C and add 10 pmol of each primer (5 μL forward and 5 μL reverse) to the boiling water. Incubate for 5 s, then turn off the heat source and allow the water to cool gradually to room temperature. Once cooled, transfer the primers to ice for immediate use.
3. Digest the pENTR4-sgRNA plasmid with BtgZI or BsaI:
4. Incubate the reaction at 37 °C for 1 h. Confirm digestion by gel electrophoresis and extract the ~3.3 kb fragment from the gel.
5. To ligate the annealed sgRNA adapters into the digested pENTR4-sgRNA vector, set up the following ligation reaction:

Component	**Volume**
pENTR4-sgRNA vector	1 μg
BtgZI or BsaI	1 μL
NEB CutSmart Buffer	2 μL
ddH_2O	to 20 μL

Component	**Volume**
Digested vector	50 ng
Annealed sgRNA adapters	1 μL
T4 DNA Ligase	1 μL
ATP (10 mM)	1 μL
T4 Ligase Buffer	1 μL
ddH_2O	to 10 μL

6. Incubate at 16 °C overnight or at 25 °C for 1 h.
7. Transform the ligation mixture into *E. coli* strain *Top10F'* and select transformants on kanamycin-containing plates.
8. Validation by colony PCR and sequencing for target insert confirm by sequencing.

Primer	Sequence
U6P-F1	5'-AAGAACGAACTAAGCCGGAC-3'
pENTR4-R	5'-CAGAGCTGCAGCTGGATG-3'

3.1.4 Gateway LR Cloning into pDEST-Cas9

1. Linearize validated pENTR4-sgRNA with ApaI or AatII; gel-purify the 2.2 kb fragment.
2. Prepare gateway LR reaction:
3. Incubate at 25 °C for 2 h.
4. Transform into Top10'F and select on kanamycin.
5. Verify correct insertion by restriction digestion:
   - For BtgZI clones: expected bands are 10 kb, 3.5 kb, and 2.8 kb.
   - For BsaI clones: expected bands are 10 kb, 3.5 kb, 2.4 kb, and 0.4 kb.

Component	Volume
pDEST-Cas9 (150 ng/μL)	1 μL
sgRNA entry vector (2.2 kb fragment)	1 μL
LR Clonase II enzyme mix	1 μL
5x LR clonase reaction buffer	
TEB buffer pH 8.0	To 10 μL

The schematic map of vector pDEST-Cas9 is given in Fig. 3.

3.1.5 Preparation of Agrobacterium tumefaciens

1. Introduce the CRISPR/Cas9 binary vector into *A. tumefaciens* strain EHA105 by electroporation or freeze–thaw method.
2. Plate the transformed *Agrobacterium* onto LB agar plates containing appropriate antibiotics (e.g., rifampicin and kanamycin) and incubate at 28 °C for 2 days.
3. Inoculate a single colony into 5 mL of LB broth with antibiotics and grow overnight at 28 °C with shaking (200 rpm).

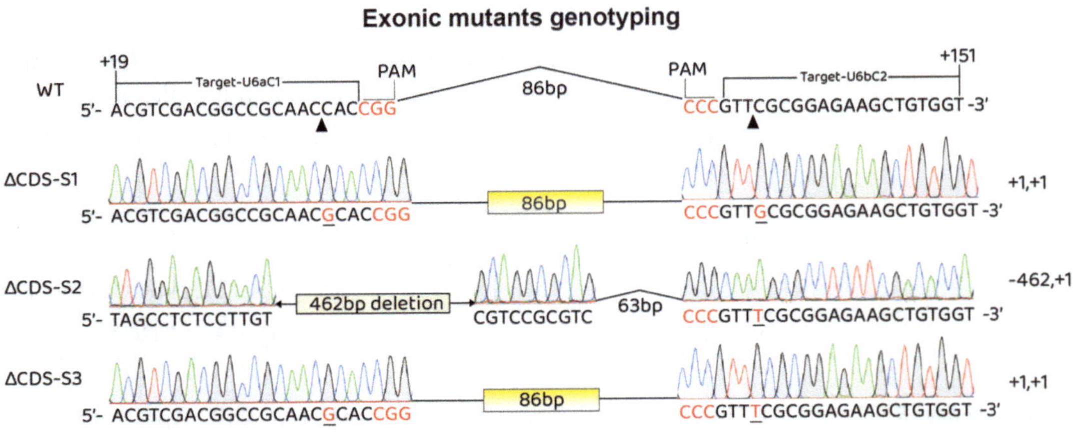

Fig. 3 The vector contains a maize ubiquitin promoter-driven rice codon-optimized Cas9 (Cas9p), a Gateway® cassette (attR1–ccdB–attR2) for sgRNA insertion, and dual selection markers—kanamycin for bacteria and hygromycin for plants

3.1.6 Pre-culture of Rice Callus

1. Use mature seeds of rice cultivar Nipponbare to initiate callus.
2. Surface-sterilize seeds and incubate them on N6D medium (supplemented with 2,4-D) in the dark at 28 °C for 7–10 days to induce callus formation.
3. Subculture friable calli onto fresh N6D medium 3–4 days before infection.

3.1.7 Infection and Co-cultivation

1. Dilute the overnight *A. tumefaciens* culture to an OD_{600} of 0.2–0.4 using AAM medium (Acetosyringone activation medium).
2. Add acetosyringone to a final concentration of 100 μM to enhance vir gene expression.
3. Immerse rice calli in the bacterial suspension and incubate for 15–30 min with occasional gentle shaking.
4. Blot excess bacteria using sterile filter paper.
5. Transfer calli onto co-cultivation medium (AAM solid medium with acetosyringone) and incubate in the dark at 22–25 °C for 2–3 days.

3.1.8 Selection and Regeneration

1. Transfer co-cultivated calli onto selection medium (N6D medium with hygromycin) and incubate in the dark for 2–3 weeks.
2. Subculture resistant calli every 10–14 days to fresh selection medium.
3. Transfer hygromycin-resistant calli to regeneration medium under light conditions (16 h light/8 h dark) to induce shoot formation.

4. Once shoots appear, transfer to rooting medium.
5. Acclimatization and Growth: Transplant rooted plantlets into soil and acclimatize under greenhouse or controlled growth conditions.

3.1.9 Mutation Validation in Transgenic Rice Plants

1. Extract genomic DNA from young leaves of T_0 plants using the CTAB method.
2. Design locus-specific primers flanking the CRISPR/Cas9 target site (200–400 bp region).
3. Perform PCR amplification using a high-fidelity DNA polymerase to ensure accurate amplification.
4. Purify PCR amplicons using a PCR purification or gel extraction kit.
5. Sequence the purified products using the locus-specific primers.
6. Analyze sequencing chromatograms using the DSDecode web tool (http://skl.scau.edu.cn/dsdecode/) [13].
7. Identify mutation types, including the following:
   - Homozygous mutants.
   - Biallelic mutants.
   - Heterozygous mutants.
   - Wild-type plants.
8. Screen plants by PCR using primers specific for the T-DNA region (e.g., hptII or Cas9 genes).
9. Plants showing successful editing at the target site, but lacking T-DNA-specific bands, are considered T-DNA-free edited plants.
10. Include appropriate wild-type and positive controls during screening.
11. Grow T_1 progeny derived from confirmed T_0 plants.
12. Extract DNA from T_1 individuals and sequence the target region to verify mutation inheritance.
13. Select homozygous, T-DNA-free plants for further phenotypic analysis or field trials.

An example of mutation identification using the above methodology is given in Fig. 4 [10].

3.2 CRISPR/Cas9-Mediated Targeted Insertion via Homology-Directed Repair (HDR)

Precise insertion of synthetic DNA sequences at specific genomic loci can be achieved through the homology-directed repair (HDR) pathway following CRISPR/Cas9-induced double-strand breaks (DSBs). This method requires two separate plasmids: (i) a CRISPR/Cas9 vector generating the DSBs, and (ii) a donor repair

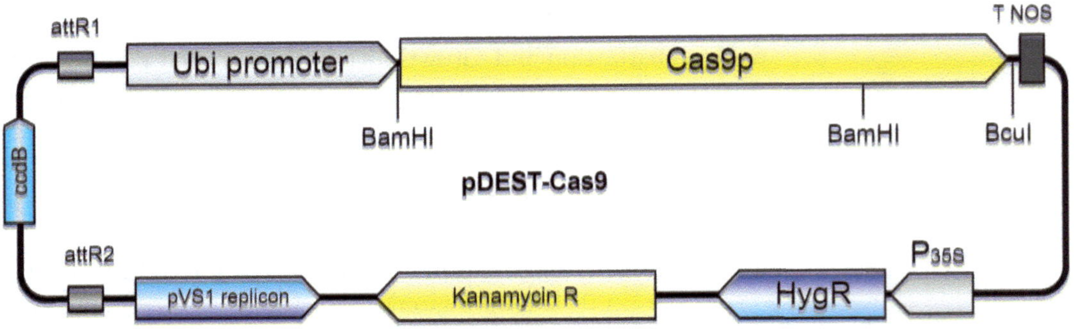

Fig. 4 Genotyping of CRISPR/Cas9-induced exon 1 mutants of *OsSULTR3;6*. Sanger sequencing chromatograms of wild-type and mutant lines (ΔCDS-S1, S2, S3) targeting exon 1. Target sites (U6aC1 and U6bC2) and PAM sequences are indicated. Mutations include small insertions (+1) and a large deletion (−462 bp), confirmed by sequencing and alignment

vector carrying the desired DNA fragment flanked by homologous arms. The following protocol, adapted from Wei et al. (2021) [12], outlines the targeted insertion of a synthetic effector binding element (EBE) into the rice *xa23* promoter.

3.2.1 Construction of Donor Repair Vector (pDxa23)

1. Select the sgRNA target sites flanking the desired insertion site within the *xa23* promoter (*see* **Note 1**).
2. Design ~200–350 bp left homology arm (LHA) upstream and right homology arm (RHA) downstream of the cleavage site (Fig. 5a) (*see* **Note 2**).
3. Insert a synthetic EBE fragment between the LHA and RHA.
4. Optionally, introduce a unique restriction site (e.g., BamHI) inside the inserted fragment for later genotyping.
5. Amplify the LHA and RHA from Nipponbare rice genomic DNA using high-fidelity PCR.
6. Verify correct size and integrity by gel electrophoresis.
7. Sequentially clone LHA, synthetic EBE fragment, and RHA into a binary backbone using EcoRI and HindIII restriction sites. Schematic diagram form [12] is shown in Fig. 5c.
8. Validate the correct assembly by colony PCR and Sanger sequencing.

3.2.2 Preparation of CRISPR/Cas9 Vector (pCxa23)

1. The CRISPR/Cas9 binary vector (pCxa23) contains Cas9 and a sgRNA designed to induce DSBs at the *xa23* promoter (Fig. 5b).
2. Confirm construction of pCxa23 by restriction digestion and Sanger sequencing.

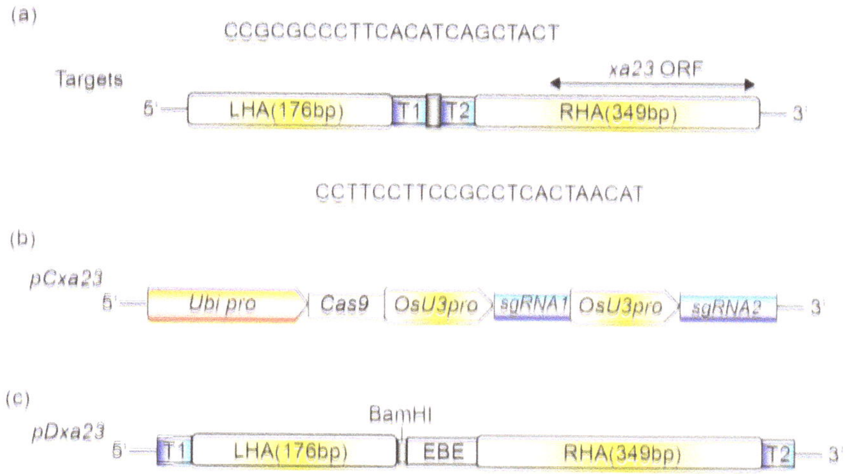

Fig. 5 CRISPR/Cas9-mediated targeted insertion of a synthetic EBE into the *xa23* promoter via HDR. (**a**) Schematic of the target region in the *xa23* promoter flanked by left (LHA, 176 bp) and right (RHA, 349 bp) homology arms. T1 and T2 indicate CRISPR/Cas9 cleavage sites flanking the insertion site. (**b**) Structure of the CRISPR/Cas9 binary vector *pCxa23*, containing Cas9 driven by the maize ubiquitin promoter and two sgRNAs under the control of OsU3 promoters. (**c**) Structure of the donor vector *pDxa23*, carrying LHA and RHA flanking a synthetic effector binding element (EBE) and a BamHI site for genotyping

3.2.3 Co-delivery into Rice Callus

1. Mix the CRISPR/Cas9 vector (pCxa23) and the donor repair vector (pDxa23) in equimolar amounts.
2. Co-deliver both plasmids into embryogenic rice calli by particle bombardment method.
3. Select transformed calli on hygromycin-containing regeneration medium.
4. Regenerate shoots and roots from hygromycin-resistant calli following the same regeneration protocols described earlier.

3.2.4 Screening and Validation of HDR Events

1. Extract genomic DNA from regenerated T_0 plants using the CTAB method (*see* **Note 3**).
2. Design primer pairs flanking the LHA–insertion–RHA junctions.
3. Perform PCR amplification to detect the presence of HDR-mediated targeted insertion.
4. Sequence positive PCR amplicons to confirm precise insertion of the EBE sequence at the target site.
5. Screen for the absence of Cas9 and selectable marker genes using specific primers.
6. Plants that carry the desired targeted insertion but lack T-DNA sequences are identified as clean edited lines.
7. Grow T_1 generation plants from selected T_0 lines.
8. Confirm the stable inheritance of the insertion and absence of T-DNA through PCR and sequencing.

4 Notes

1. Dual sgRNAs creating DSBs flanking the insertion site can enhance targeting efficiency.
2. Homology arms of 200–350 bp yield optimal HDR efficiency.
3. HDR-based insertion events are typically rare (~1–3%); therefore, screening a large number of T_0 plants is necessary.

References

1. Ristaino JB, Anderson PK, Bebber DP, Brauman KA, Cunniffe NJ, Fedoroff NV et al (2021) The persistent threat of emerging plant disease pandemics to global food security. Proc Natl Acad Sci USA 118:e2022239118
2. van Schie CC, Takken FL (2014) Susceptibility genes 101: how to be a good host. Annu Rev Phytopathol 52:551–581
3. Savary S, Willocquet L, Pethybridge S, Esker P, McRoberts N, Nelson A (2019) The global burden of pathogens and pests on major food crops. Nat Ecol Evol 3:430–439
4. Oliva R, Ji C, Atienza-Grande G, Huguet-Tapia JC, Perez-Quintero A, Li T et al (2019) Broad-spectrum resistance to bacterial blight in rice using genome editing. Nat Biotechnol 37:1344–1350
5. Li S, Lin D, Zhang Y, Deng M, Chen Y, Lv B et al (2022) Genome-edited powdery mildew resistance in wheat without growth penalties. Nature 602:455–460
6. Wang M, Li S, Li H, Song C, Xie W, Zuo S et al (2023) Genome editing of a dominant resistance gene for broad-spectrum resistance to bacterial diseases in rice without growth penalty. Plant Biotechnol J 22:529–531
7. Wang F, Wang C, Liu P, Lei C, Hao W et al (2016) Enhanced rice blast resistance by CRISPR/Cas9-targeted mutagenesis of the ERF transcription factor gene OsERF922. PLoS One 11:e0154027
8. Schiermeyer A, Schneider K, Kirchhoff J, Schmelter T, Koch N, Jiang K et al (2019) Targeted insertion of large DNA sequences by homology-directed repair or non-homologous end joining in engineered tobacco BY-2 cells using designed zinc finger nucleases. Plant Direct 3:e00153
9. Marchal C, Pai H, Kamoun S, Kourelis J (2022) Emerging principles in the design of bioengineered made-to-order plant immune receptors. Curr Opin Plant Biol 70:102311
10. Shafique MS, Yapei L, Man L, Hongjie W, Ruyi S, Chunlian W, Zhiyuan J (2024) Coevolution unveiled: sulfate transporters mediate rice resistance and susceptibility to Xanthomonas oryzae pv. oryzicola. Plant Biotechnol J 22:2632–2634
11. Xu X, Xu Z, Li Z, Zakria M, Zou L, Chen G (2021) Increasing resistance to bacterial leaf streak in rice by editing the promoter of susceptibility gene OsSULRT3;6. Plant Biotechnol J 19:1101–1103
12. Wei Z, Abdelrahman M, Gao Y, Ji Z, Mishra R, Sun H et al (2021) Engineering broad-spectrum resistance to bacterial blight by CRISPR-Cas9-mediated precise homology directed repair in rice. Mol Plant 14:1215–1218
13. Liu W, Xie X, Ma X, Li J, Chen J, Liu YG (2015) DSDecode: a web-based tool for decoding of sequencing chromatograms for genotyping of targeted mutations. Mol Plant 8:1431–1433

INDEX

Tsutomu Kawasaki and Hirofumi Yoshioka (eds.), *Plant Immunity: Methods and Protocols*, Methods in Molecular Biology, vol. 3012, https://doi.org/10.1007/978-1-0716-5138-4,